Smart Cities in Application

Stan McClellan

Editor

Smart Cities in Application

Healthcare, Policy, and Innovation

 Springer

Editor
Stan McClellan
Ingram School of Engineering
Texas State University
San Marcos, TX, USA

ISBN 978-3-030-19398-0   ISBN 978-3-030-19396-6   (eBook)
https://doi.org/10.1007/978-3-030-19396-6

This Springer imprint is published by the registered company Springer Nature Switzerland AG
The registered company address is: Gewerbestrasse 11, 6330 Cham, Switzerland

Foreword to "Smart Cities in Application"

"Please Text, Don't Drive"

As funny as that sounds, it was the statement I made at the Regional Transportation Council, the independent transportation policy body of the North Texas Metropolitan Planning Organization. We had just received an update on autonomous vehicles from the National Highway Traffic Safety Administration (NHTSA). I was partly trying to be funny, but also serious at the same time. When we reach full "level six automation", the consensus is that will actually be safer to text while you drive than driving your vehicle yourself! So what is our future really going to bring?

As the Former Mayor of Frisco, Texas, a board member for Collin College Educational Foundation, a commissioner for the State of Texas Red River Boundary Commission, and a CEO of the Prosper Economic Development Corporation, it has been an honor and privilege to serve my local community during complex transitions and astounding growth. I have witnessed the implementation of many incredibly valuable technologies, applications, and concepts which revolutionize the management of municipal facilities and assist the educators, public safety officials, and city management in improving the lives of the citizens and neighbors. In my current role as a Principal at Ryan, LLC, I advise cities, counties, and states on best practices, and I represent companies as they look for sites in this thriving new frontier of "Smart Cities."

You will see a focus on Smart Cities in this book, but what does that really mean? While I won't presume to define it for you, I can tell you that my definition includes thinking outside the box of what we consider "norms." A truly Smart City uses innovation to provide critical services to the residents in a cost-effective manner. Issues such as public policy; equitable treatment for citizens; effective technologies for hospitals, police, and fire; and other public safety concerns are paramount. Intelligent transportation systems which are integrated with command/control will be critical assets for traffic optimization, event management, and orchestration of high-priority travelers, such as first responders. Internet-of-Things (IoT) architectures, cross-domain enterprise integration of devices, and software and management systems will enable new solutions to common problems. Smart home, smart building, smart grid, and smart water solutions will optimize limited resources and provide equitable, inclusive environments for the residents.

Frisco, Texas

As the Mayor of Frisco, Texas, from 2008 to 2017, I was privileged to be able to experience firsthand some of the policy issues and incredible innovations that lead to a "Smart City." It was also exciting to be on the front end of the amazing opportunities that exist in this space. Frisco has experienced incredible growth in the last 25 years, from just over 6000 residents in 1990 to almost 180,000 in 2017. Managing this "hyper-growth" required a lot of innovation, an attention to detail, and a vision of the future.

As a case in point, the city of Frisco is in Texas, where water resources are limited. We addressed this problem in 2001 by being the first city in the United States to adopt Energy Star standards for new homes. The innovative "Smart Controller" program was implemented in 2007, so that each new home is connected to a weather station to predict weather conditions. As a result, grass, plants, and other household items use water only when needed. Innovation, through careful management of resources, bold leadership, and matching technologies to requirements, was the key to achieve a level of "smartness" that helps everyone.

Innovation Is Key

In a "Smart City," innovation will be critical in deploying and optimizing municipal services. This certainly includes new technologies and systems which support public safety. The Frisco Independent School District, which grew from 6 schools in 1996 to 72 schools in 2018, had the opportunity to partner with the city and security providers to create a system that gives first responders an access to maps and live data feeds on calls to schools and an up-to-date access to school building information—all in real-time. These smarter and safer communities will continue to create an atmosphere for families and corporations to achieve success.

In a "Smart City," innovation will be critical in developing new approaches to healthcare. Do we continue just treating the symptoms or use innovative practices? Do we use an app to summon a health practitioner to our home to treat the typical ear infection within a short period, or do we continue routine doctor's office visits? With an innovative approach, the doctor may use augmented/virtual reality (AR/VR) to treat patients remotely, and the first responders may have access to novel training and situational awareness tools which improve their performance. Technology deployed in appropriate contexts can help ensure the safety of the first responders as well as the safety of those in their care. With appropriate innovation in healthcare, patients will experience completely personalized care, and the result will include better patient outcomes and commitment to quality and safety.

When Jerry Jones and I were sitting in the Cowboys bus at the grand opening of The Star sports fields in Frisco, the subject of innovation and the future of sports medicine and research on how to prevent sport injuries were the topics we were exploring. Jump forward a few short years and now, The Star in Frisco integrates healthcare into the recreation facility. In one place, you have an amateur, club, semi-pro, and professional sports and the opportunity to marry innovative research in injuries, medicine, rehab, nutrition, and many other concepts. This is an example of innovation which results in better training, personalized healthcare, and improvements in equitable treatment for all citizens of the community.

Enabling all of these innovations presupposes a minimum level of universal coverage, an ability to ensure commonality of understanding, and a means of providing a trusted service. Implementing a vision of this nature will set apart the truly "Smart City" from the ranks of conventional municipalities. Innovations, such as block chain technology, may be important in the authentication of transactions in the next-generation power distribution. It may also be critical in ridesharing markets enabled by autonomous vehicles. How will cities manage these digital transactions? New technology deployments, including fifth-generation telecommunications or "5G," will provide a baseline of connectivity to support high-priority, low-latency, and mission-critical municipal services as well as entertainment and social interaction for citizens, businesses, and agencies. How will cities manage the public/private partnerships and leverage of municipal assets to ensure "fairness" across citizen groups?

Challenges and Constraints

All of these technologies will create opportunities for automation of complex processes (e.g., driving, delivery, response) as well as issues of digital trust and security (e.g., encryption, authentication) and policy-based rationale for centralized management as well as distributed intelligence. The chapters in this book outline approaches and recommendations for architecture/implementation, technology choices/constraints, and considerations for this new paradigm which will drive significant changes across all aspects of society and vertical segments in the "Smart City." Along this evolutionary—and revolutionary—path, the ability to evaluate new ideas, optimize innovation, forecast skill sets, and prioritize markets will be the key areas of innovation in the Smart City space.

Smart Cities and innovation also bring a host of challenges that society must help answer. No one can argue that sending a drone automatically over an area where a 911 call was received helps the first responders become more situationally aware prior to arrive at the scene. However, what about our expectation of privacy? With automated traffic lights that capture video, who will have access to that data? The one answer that is clear is that innovation will continue its break-neck speed and we must all learn to work and function in this new world!

Looking Forward

By the time you read this book, the National Soccer Hall of Fame museum at Toyota Stadium in Frisco will be open. With state-of-the art technology and facial recognition by the NEC, you can join your favorite team on the field and visit with your favorite legends. How we enjoy entertainment is changing just as rapidly. A few years ago, no one had heard of eSports. Now, almost every professional sports team is involved. Twitch, OpTic Gaming, NRG, complexity Gaming, and Overwatch League are names you may have never heard. Over the next few years, they will become more recognizable. The US Conference of Mayors has asked me to help the Mayors from around the country understand what a public/private partnership looks like in eSports. The first thing I told them is that no one really knows where it is going, but everyone wants a foot in the door. So, you are getting your foot in the door with this book!

I hope that you will enjoy the unique perspectives on Smart Cities that this book provides. It is a continued perspective on Smart Cities, which builds on the companion book *Smart Cities: Applications, Technologies, Standards, and Driving Factors* (2018). This top-down exploration of the Smart City phenomenon is targeted toward the practitioners, and the conversational nature of the book will appeal to anyone interested in learning more about key issues in this important new space.

Maher Maso, Principal – Credits and Incentives, Ryan, LLC
CEO—Prosper Economic Development Corporation (2017–2018)
Commissioner—State of Texas Red River Boundary Commission (2013–present)
Board Member—Collin College Educational Foundation (2006–present)
Mayor—Frisco Texas (2008–2017)

Preface

This book provides a continued perspective on the Smart City phenomenon, which builds on the companion book *Smart Cities: Applications, Technologies, Standards, and Driving Factors.*

The top-down exploration of the Smart City phenomenon continues in this volume with a focus on topical areas of "smart healthcare," "public safety and policy issues," and "science, technology, and innovation." The contributors with direct and substantive experience with the important aspects of Smart Cities discuss issues with technologies and applications, roadblocks to implementation, innovations which create new opportunities, and other factors relevant to emerging Smart City infrastructures.

The continued exploration of Smart City issues, via a sequence of editions in topical areas, is unique in the marketplace and is targeted toward the practitioners in various fields. Additionally, the practical and "conversational" nature of the coverage is appealing to the readers interested in learning more about Smart City issues and technologies.

San Marcos, TX, USA

Stan McClellan

Contents

Part III Science, Technology, and Innovation

About the Authors

Michelle Alvarado is an Assistant Professor in the Department of Industrial and Systems Engineering at the University of Florida, where her research develops theory, models, algorithms, and practical tools to improve the operation of complex healthcare systems under uncertainty.

Alvarado's methodological expertise is in simulation and stochastic programming. Her current research interests include hybrid simulation modeling for healthcare operations, remote health monitoring of chronic diseases, and optimal design of health policy models. She is the Co-founder and Co-director of the HEALTH-Engine Laboratory at the University of Florida. She received her MS and PhD degrees in Industrial Engineering from Texas A&M University (TAMU), in 2012 and 2014. Her research has been published in *Health Care Management Science*, *Journal of Medical Internet Research*, and *IISE Transactions on Healthcare Systems Engineering*, *IEEE Transactions on Automatic Science and Engineering*, and *Simulation*. She is a Member of IISE, INFORMS, and ASEE.

Achamkulamgara Arun is a Connected Products Director for Cognizant Technology Solutions (CTSH), a $14B company headquartered in New Jersey, NJ, with over 100+ delivery centers around the world. He serves as Chief Architect for Internet of Things (IoT), Cognizant Connected Products, a part of Cognizant's Digital Enterprise with focus on developing product solutions.

Arun develops solution architectures, product design, and development plans for customers across smart cities and facilities, manufacturing, medical devices, healthcare, retail, and communication products industries. His interests are in developing standardized microservices-based software and embedded solutions, device security, and product compliance. Prior to Cognizant, he served as Technical Director and Senior R&D Manager, developing and supporting enterprise level platform, hardware, and software worldwide and working at Intel, 3M, Singapore Technologies, and Central Research Labs. He holds an MS from Indian Institute of Technology, India, and an MBA from the Carey School of Business at Arizona State University.

Meredith Barrett, PhD is the Vice President of Science & Research at Propeller Health, a digital health company dedicated to improve outcomes in respiratory disease, including asthma and COPD.

She works to leverage Propeller's sensor technology to generate insights about respiratory disease for patients, providers, and communities. Her training in ecology, population health, and spatial analysis has enabled her to study the impacts of environmental change on both infectious and chronic diseases. She was a Robert Wood Johnson Foundation Health & Society Postdoctoral Scholar at the University of California Berkeley School of Public Health and UC San Francisco Center for Health and Community. She completed her PhD in Ecology with a focus on global health at Duke University, where she was a National Science Foundation Graduate Research Fellow. Her research has been published in *Science*; *Environmental Health Perspectives*; *Frontiers in Ecology and the Environment*; *Annals of Allergy, Asthma & Immunology*; *Journal of Big Data*; and *Preventing Chronic Disease*, among other journals and in popular media, such as *The Huffington Post*.

Catherine Crago Blanton is the Head of Strategic Initiatives and Resource Development in the Housing Authority for the City of Austin, TX (HACA).

At HACA, she played an active role in building the Unlocking the Connection program, a first-of-its-kind initiative to bring Internet connectivity, digital literacy, and devices to every public housing authority resident. The program was the inspiration for the White House-HUD initiative called ConnectHome. For more than 10 years, she helped the state agencies and local governments assess the business case for digital inclusion, to enhance revenue, to support emergency preparedness, and to promote equity in diverse communities.

Crago Blanton has presented at the White House and has provided testimony to the Federal Communications Commission on the impact of connectivity in low-income populations. She serves on the Board of Advisors and on the advisory boards of SXSW Interactive Festival and lectures in the University of Texas Master's in the Human Dimensions of Organizations. She also serves as Vice President of Diversity, Equity, and Inclusion of the Board of Directors of Austin CityUP, where she also co-chairs the Housing Committee, and is a Recipient of the Texas Diversity Council DiversityFIRST Award for work in diversity and inclusion.

Mike Brown is an Institute Engineer for Southwest Research Institute (SwRI). He has been a Leader in the Development of Intelligent Systems for over 22 years, serving various federal, state, and commercial clients in projects spanning the areas of advanced traffic management and traveler information systems, connected and automated vehicles, and Smart Cities. He currently serves on the Board of Directors for the OmniAir Consortium, an industry association promoting interoperability and certification for connected vehicles, intelligent transportation systems (ITS), and transportation payment systems.

As an Institute Engineer at SwRI, Brown provides expert consultation services for SwRI programs and tackles highly specialized problems on behalf of the clients. He also works with Senior SwRI and Division Staff to plan future technology needs and to lead the development of new programs in his areas of expertise. He also serves as a Member of SwRI's Advisory Committee for Research and is a Subject Matter Expert for numerous standards committees in standards development organizations, including the Institute of Electrical and Electronics Engineers (IEEE), Society of Automotive Engineers (SAE), and International Organization for Standardization (ISO).

Adam Cason Director of Product Marketing at Futurex, is responsible for the company's virtually driven content marketing initiatives, technical documentation portfolio, and engagement strategy for customer and partner relationships. In this role, he works closely with executive audiences around the world, focusing on the growth of Futurex's thought leadership within the overall field of cybersecurity.

George Koutitas is an Entrepreneur and Academic in Electrical & Computer Engineering. He has more than 10 years of business and academic experience in smart grids, Internet of Things, augmented reality, and wireless networks.

Koutitas is an Assistant Professor of Electrical and Computer Engineering in the Ingram School of Engineering at Texas State University, is the Director of the XReality Research Lab where he works on the intersection of AR and IoT, and is also the Author of one book, two book chapters, and two patent applications. He has published more than 47 scientific publications in peer-reviewed journals and conferences. His research work has received more than 990 citations.

In 2014, Koutitas founded Gridmates (www.gridmates.com) which is the world's first cloud platform for Smart Energy Donations helping electric utilities and retailers transform their bill assistance and social responsibility programs with digital innovation. In 2019, Koutitas co-founded Augmented Training Systems Inc. (http://augmentedtrainingsystems.com), a software company which develops training platforms to help first responders improve performance while reducing training costs.

Grayson Lawrence has over 18 years of graphic design and design thinking experience, specializing in user interface (UI), user experience (UX), game, Web, virtual reality (VR), augmented reality (AR), and mobile application design.

As part of his research, he has developed award-winning designs for mobile applications, such as DocbookMD, an application designed to foster better communication between doctors and their care teams, and Govely, an application designed to encourage civic engagement among high school students. In addition, he has collaborated on interdisciplinary teams for a variety of other mobile and VR applications focused on mental health, including alcohol abuse, veteran students with PTSD, emergency medical technician (EMT) training, and medication adherence via voice UI and mobile application/wearable integration.

Joe Moorman is a Software Engineer who has worked in C++/web systems development with the Reynolds and Reynolds Company in Houston, TX, through 2017 and, more recently, in financial technology consulting with Riskcare Inc. in New York, NY. The areas of financial industry research included short-term equity market impact models using multivariate Hawkes processes and continuous calibration, for which he implemented a C++/Python software framework. Presently, he is a Full-Lifecycle Software Engineer with Myriad RBM, Inc., of Austin, TX, in the biotechnology space, automating and streamlining quality control processes, and workflow tasks to improve immunoassay processing times and quality for a CLIA-certified laboratory.

Keith Noble is a Commander with Austin-Travis County Emergency Medical Services (EMS). He is currently assigned to the department's Homeland Security and Emergency Management Division focusing on integrating EMS into disaster, mass causality, homeland security, and emergency management plans for the City of Austin. He has been involved with EMS for over 12 years and some type of public safety for 15 years.

Noble is also responsible for one of the 13 "Am-Bus" emergency vehicles in the State of Texas, which provides full advanced life support care for large-scale incidents. Each Am-Bus provides facilities equivalent to six ambulances, including space for 20 patients. These emergency preparedness solutions consolidate resources and reduce wait times for treatment during large-scale or mass-casualty incidents.

He obtained his master's degree in Emergency Management from Jacksonville State University and is currently working on a Master's in Public Administration. He also holds a BS degree in Criminal Justice from The Pennsylvania State University.

Steve Pearson is the Lead Strategist and Founder of The Pearson Strategy Group, LLC, which has clients including legal teams, companies too small for dedicated R&D departments, inventors, and innovators across software and technology fields.

He believes that solid research is one of the most valuable tools for accelerating innovation and product development. He has staked his business reputation on knowing how and where to dig into key information to provide this competitive advantage for entrepreneurs and companies. Clients also rely on his expertise to help interpret trends, unknown competitors, and strategies in the revealed data. By delivering targeted research based on patents and markets, along with strategic consulting, he empowers clients to move forward quickly and at a minimal cost.

Pearson is an electrical engineer with experience in semiconductor manufacturing, machining, business analysis, and patent research and served aboard a Navy submarine as a Nuclear Reactor Operator. He holds several patents, with intimate knowledge of all aspects of the patent process. He is a Past Chairman of the Austin chapter of the IEEE's Power and Energy Society. He is happily married to a career coach, and they have two outgoing children and three very friendly dogs.

Eduardo Pérez-Roman is an Associate Professor in the Ingram School of Engineering at Texas State University. He received his PhD in Industrial and Systems Engineering from Texas A&M University in 2010 and his BS in Industrial Engineering from the University of Puerto Rico at Mayagüez, Puerto Rico, in 2004.

He was a Postdoctoral Research Associate in the Department of Industrial and Systems Engineering at Texas A&M University from 2010 to 2012.

Pérez' research interests are in the use of methodologies and theories in operations research, systems engineering, discrete-event simulation, algorithms and software design, and decision theory analysis to solve problems in healthcare and renewable energy. He is the Director of the Integrated Modeling and Optimization for Service Systems (iMOSS) research laboratory. His research has been sponsored by the National Science Foundation (NSF), Robert Wood Johnson Foundation, Baylor Scott & White Health System, Adventist Health System, and NEC Corporation, with results published in multiple journals, including *IIE Transactions*, *IIE Transactions on Healthcare Systems Engineering*, *Simulation*, Health care Management Science, and *Computers & Operations Research*.

Liam Quinn is a Sr. Vice President and Sr. Fellow at Dell Technologies. He has been with Dell since 1997 and currently leads the Pan-Dell Technologies 5G strategy. His key areas of focus and research include converged mobility, digital and workforce transformation, security and manageability, Internet of Things (IoT), and specific applications of augmented and virtual reality (AR/VR) in areas of remote maintenance, gaming, and 3D applications.

Quinn has over 120 granted and pending US patents and is a published Author on networking. He was named Dell Inventor of the Year in 2005, 2007, and 2014. He is a frequent Speaker at Wi-Fi and Networking Forums and also represents Dell on the boards of the Wi-Fi Alliance and the OpenFog Consortium. He has designed network systems and wireless solutions and has managed numerous engineering teams in systems architecture and product development. He serves on the Advisory Board of the University of Texas, Cockrell School of Engineering, and the Texas State Ingram School of Engineering. He holds undergraduate and graduate degrees in Electrical Engineering and Computer Engineering.

Scott Smith is the co-founder and CEO of Augmented Training Systems (ATS), a software company which develops training platforms to help first responders improve performance while reducing training costs (http://augmentedtrainingsystems.com). ATS builds custom, human-centered augmented, and virtual reality training simulators for equipment and processes that are cost-prohibitive or dangerous. Their training scenarios can be deployed anywhere and anytime to multiple trainees, saving time and money and improving situational awareness for the whole team.

Smith received a PhD in Educational Measurement and Statistics, Educational Psychology, and Learning Systems from Florida State University (PhD) and has over 10 years of clinical experience working with high-risk youth, substance abuse and addiction, families, trauma, domestic violence, and the homeless population. He is the Director of the Virtual Reality and Technology Lab, which develops treatments for returning veterans experiencing social anxiety, and an Associate Professor of Social Work at Texas State University.

Michael Stricklen is an Executive Director in the EY-Parthenon practice of Ernst & Young LLP, based in Boston, and is a Member of its Software Strategy Group. He provides deep knowledge in product and technology diligence projects and leverages 20 years of experience in software engineering and enterprise architecture to optimize software products' technology stacks, creating efficiencies, maximizing functionality, and minimizing infrastructure performance risks.

Stricklen's experience covers Executive, Entrepreneur, Board Member, and Advisor positions with high-tech companies. He has worked within and acted as an Advisor to a broad range of software and software-driven companies, including enterprise systems and data center management; healthcare management software; telecommunications and high-speed networking; integrated financial trading systems; mobile device management; governance, risk, and compliance management; and infrastructure- and platform-as-a-service (IaaS/PaaS) architectures.

Walt Trybula is the Executive Director of the Trybula Foundation, a Texas for-profit corporation founded in 1999 (http://www.tryb.org), which works with the organizations in evaluating hi-tech and emerging technologies to determine the potential for successful inclusion in the organization's strategic direction.

With the experience ranging from startups to Fortune's top ten corporations, the foundation's business acumen melds the two worlds of business and technology to provide guidance to ensure successful programs. Over the last 25 years, activities have included evaluation of the economic impact of technology on existing industry cost structures, reports on the economic impact of technology on cities and surrounding areas, and evaluation of emerging technology on corporate expansion.

Trybula is a Fellow of the IEEE and of the SPIE and was a Senior Fellow at SEMATECH where he was responsible for leading the initial effort to evaluate the feasibility immersion lithography. He has an undergraduate degree in Physics from the Illinois Institute of Technology, an MBA from James Madison University, and a PhD in Information Sciences from the University of Texas at Austin.

David Wierschem is the Associate Dean for Undergraduate Programs in the McCoy College of Business Administration at Texas State University. His academic experience includes the positon of Chair of the Department of Computer Information Systems and Quantitative Methods as well as teaching Data Communications, IT Security, Statistics, and numerous other technology-related courses.

Wierschem received his MBA from Georgia State University and his MS in Operations Research and PhD in Management Information Systems from the University of Texas at Dallas. He has written numerous articles in the areas of IT productivity, data warehousing, and IT employment. Currently, his research interests include IT employment and the use of motion capture technology to analyze human safety and efficiency in the areas of manufacturing and materials handling.

Part I
Smart Healthcare

Personalizing Healthcare in Smart Cities

Eduardo Pérez-Roman, Michelle Alvarado, and Meredith Barrett

Contents

1 Introduction

Healthcare personalization is a new opportunity for health care organizations in their pursuit of better outcomes in their service provision process [1]. Healthcare personalization focuses on patient-centered health care, personalized health planning, and patient engagement. Much of previous research in this area has viewed service variability as something bad that must be limited. However, patient heterogeneity with variability in service needs provides an opportunity to deliver more value for patients through the personalization of services. One major gap in the

E. Pérez-Roman (✉)
Ingram School of Engineering, Texas State University, San Marcos, TX, USA
e-mail: eduardopr@txstate.edu

M. Alvarado
Industrial and Systems Engineering, University of Florida, Gainesville, FL, USA

M. Barrett
Science and Research, Propeller Health, Madison, WI, USA

literature on customization is the lack of research on service personalization. Strategies for manufacturing firms are considerably different from strategies for service firms and they need to be considered in a separate manner [2–4].

Recent research has established that service personalization is challenging mostly because it deals with two types of variability sources: service personnel and customers [5]. One of the challenges involves deciding between reducing and accommodating customer introduced variability [6]. There are five phases in which customer-based variability can occur: customer arrival, customer requests, customer capability with respect to their expected involvement with the process, effort that customers are willing to exert, and subjective preferences of how service should be provided. In addition, there is variability that is not attributable to the customers but still impact their outcomes. For example, employees can be heterogeneous in skill levels.

Research in patient healthcare personalization is limited. A smart city can help hospitals achieve better healthcare services through patient healthcare personalization. A smart city allows for the introduction of intelligent management systems that support the digital collection, processing, storage, transmission, and sharing of patient information such as personal information and social information. In addition, the infrastructure of a smart city can help in solving many health hazard problems by supporting different sections of healthcare systems including the intelligent management and supervision of health data, medical equipment and supplies, communication systems, automated management, and supervision of public health [7].

Current research on patient healthcare personalization considers the following two approaches: (1) patient health outcomes and (2) resource utilization and system cost. The patient health outcomes perspective has been applied to model the factors in definitive care that maximized patient outcomes [8–11]. Research considering resource utilization and systems cost for patient healthcare personalization is limited. Most of the available studies compare patient driven methods for scheduling appointments [12] or queuing systems in which customers had a state-dependent probability of not being served [13].

Patient healthcare personalization and smart health are two areas of research that are strongly connected and that will positively impact the future of the concepts of patient wellness and well-being [14]. Both areas of research require large volume of data. Data sources include biomedical sensors, (e.g., temperature, heart rate), genomic driven data (gene expression, sequencing data), payer–provider data (pharmacy prescription, insurance records), and social media data (patients' status, feedback) actuators, to observe and to predict the best course of action to improve patients' outcomes. The rest of the chapter is organized as follows. Section 2 provides an overview of personalized healthcare within the context of smart cities. Section 3 presents a case study that combines the use of smart cities and patient personalization concepts in healthcare. Section 4 presents challenges for personalized healthcare within the context of smart cities and Sect. 5 provides a summary of the chapter with concluding remarks.

2 Overview of Personalized Healthcare Within the Context of Smart Cities

In the last two decades, a number of healthcare systems have been introduced including digital healthcare systems (software and/or internet-connected devices), electronic health record systems, hospital-based systems, and finally smart healthcare system. Through investigating the relevant literature, we present four areas within the healthcare setting where personalized healthcare within the context of smart cities is applied. The four areas are patient scheduling and resource planning, management of healthcare associated infections, remote technologies, and patient treatments and diagnosis.

2.1 Patient Scheduling and Resource Planning

Scheduling systems are used by primary and specialty care clinics to manage access to healthcare resources. Different factors impact the performance of scheduling systems including patient arrivals and service time variability, preferences from patient and providers, available technology, and the experience level of the scheduling staff. In this section, we discuss advanced technology, models, and algorithms for personalized patient scheduling within the smart city context.

2.1.1 Patient Portals and Smart Rooms

Online patient portals are electronic tools to facilitate healthcare self-management support [15]. Patient portals enable online access to portions of medical records, electronic communication with medical care team, appointment reminders, viewing lab test results and images, requests for scheduling non-urgent medical appointments, requests for prescription renewals, access to patient education resources, adding or editing allergy lists, downloading or completing medical forms, and paying medical bills. An increasing number of health care organizations are using patient portals, and they have been impactful in their quality improvement outcomes and implications in meaningful use. A healthcare system typically contracts with vendor to provide the patient portal. Vendors of popular patient portals that support scheduling of appointments include Athenahealth, Cerner, Epic Systems (via MyChart), Intelichart, MEDITECH, and RelayHealth [16]. Figure 1 presents some of the devices that can be used with patient portals including cell phones, tablets, and computers. However, many limitations remain for patient portals such as the inability to transmit and share health information with a third party [17].

Smart room technology is an example of smart systems capable of transforming patient service by eliminating redundant actions and facilitating information required for decision-making. For instance, The University of Pittsburgh Medical

Fig. 1 Examples of common patient portals and user devices

Center (UPMC) has designed a smart room system capable of disseminating diverse sets of patient data to different provider categories. The room assists in the decision-making process for clinical care and also is responsible for excluding 60–70% of redundant actions related to documenting routine clinical care [18].

2.1.2 Appointment Scheduling Algorithms

Traditionally, many appointments were provided to patients on a first-come, first-serve basis. At best, appointment schedulers were using a template to fill in appointments as requests arrived. Now there are sophisticated scheduling algorithms in use to find optimal appointment times based on a set of stated objectives. These algorithms strategically schedule complex appointments that perhaps require multiple resources, have time-sensitivities or uncertain durations [19]. The clinic may wish to maximize the number of patients seen daily to increase both revenue and improve access to care. However, there are typically competing objectives such as minimizing waiting time for the patient, reducing overtime for the provider, or balancing workload among the staff [20]. The scheduling algorithms are typically customized to a particular healthcare setting, such as those that have high rates of no-shows or that require a multi-step process [21–26].

2.2 Healthcare Associated Infections

Healthcare associated infections (HAIs) remain a significant problem for healthcare providers in the USA and research studies addressing this subject have emerged over the last 10 years. Currently, there is a need for smart solutions that can help in deciding the best course of action according to different patient types and the status of their condition. For example, tools are needed to help answering the following questions: (1) which patients must be screened upon admission to prevent the development of HAIs?, (2) how to reduce microbial contamination?, and (3) how to use data on patient demographics and conditions to minimize infection rates? In this section, we discuss smart tools and methods for personalized patient services when targeting HAIs.

2.2.1 Touchless Disinfecting Technologies

Reducing microbial contamination in the healthcare environments is challenging. Healthcare associated infections has become an area of increased scrutiny after the implementation of the Affordable Care Act. Recently, hospitals are adopting touchless technologies to overcome some of the deficiencies of manual cleaning by eliminating the human element from process. Touchless technologies include multiple products including fumigation methods and self-disinfecting surfaces. While some studies have demonstrated that certain methods can eliminate experimentally placed bacterial samples, no studies were able to document complete killing of all organisms contaminating in real settings [27]. Complex surfaces and devices with different shapes in the hospital room continue to harbor pathogens that touchless methods are unable to eliminate completely.

2.2.2 Advanced Analytics and Simulation

The hospitals for sick children (SickKids) in Toronto implemented a smart method for preventing patients from acquiring life-threatening nosocomial infections. The hospital uses advanced analytics to study big amounts of data collected from patient monitoring devices. The data analysis allows to search for early detection of potential signs of infection. This new strategy provides the benefit of patient early treatment since it can diagnose potential infections 1 day earlier than the previous method [7].

Pérez et al. [28, 29] extended the current methods for addressing CAUTIs to a new level beyond the current state of practice by introducing a new discrete-event simulation model for system assessment and determination of the clinical efficacy of CAUTI preventive interventions. The simulation model provides a platform where proposed clinical interventions can be studied and analyzed; giving an idea of possible expectations before clinical research is conducted; saving time, money, and minimizing risks. The computational results provide useful insights into patient service management for CAUTI prevention and show that operational factors such as the nurse: bed ratio, catheter daily removal chance, and late maintenance risk have a significant impact in the intensive care unit CAUTI rate.

2.3 Remote Technologies

Remote health is a type of ambulatory health care that allows patients to use technology to collect data and communicate with their health care provider in a different location. Remote healthcare is possible through health information technology such as remote appointments, advanced sensors, wearable devices, and machine learning. This type of healthcare is useful for monitoring a patient's daily activity and vitals, checking compliance to prescribed treatment regimens, and communicating with providers between scheduled in-person visits. Remote healthcare can substitute for traditional care in some instances, but is primarily intended to support traditional care. Remote healthcare is especially useful for rural populations and patients with chronic conditions such as diabetes, dementia, and cardiovascular disease.

2.3.1 Wearable Devices

Smart wearable devices collect and monitor biomarkers including body temperature, blood pressure, heart rate, oxygen saturation, blood sugar levels, or walking and standing patterns. These devices are worn on the patients' body such as a watch on the wrist, a patch on the arm or body, shoes on the feet, or a strap on the hand or fingers. Medical devices including asthma inhalers and sleep apnea machines also have the ability to transmit usage patterns for providers to review in clinic. Even smart bras can evaluate breast health status for women as a means of cancer screening.

The design of the wearable devices determines how data is collected and transmitted for storage, analysis, and decision-making. Some patients may be expected to do routine (e.g., daily) manual download or upload of the data for transmission to their care provider. However, the most successful approaches rely on procedures that are minimally disruptive to the patient's normal daily routine. Automatic wireless transmissions yield better long-term use and impact [30]. Wearable device technology is rapidly advancing to become leaner, more flexible, accurate, and less burdensome to the human body to ensure increased reliability and long-term utilization.

2.3.2 Ambient Sensors

Ambient sensors capture data from the surrounding physical environment. These sensors are not worn by the patient, but instead occupy a fixed position to capture motion, pressure, video/audio, water consumption, ambient temperature, ambient light, CO_2 concentration, heat, and allergens [31]. In some situations, the patients forget to put on a wearable device or do so incorrectly. In these instances, the ambient sensors become useful for continuing to monitor the patient in their home and/or community. Research has shown that ambient sensors can identify movement and sleep patterns [32].

2.3.3 Machine Learning

Due to the rapid data collection enabled by wearable devices and ambient sensors, there is a need to figure out what to do with the abundance of incoming data. Machine learning is a method of data analysis that automates the building of models to learn from the data, identify patterns, and make decisions. In the healthcare setting, machine learning has many practical implications because the machine learns typical or normal patterns for a patient and learns how to detect a deviation [32]. Consider a patient wearing a continuous blood glucose monitor. The machine would learn typical blood sugar level patterns for the patient and detect a pending incident of hyper- or hypoglycemia when those patterns deviate from the norm [33]. Essentially, machine learning techniques are a way to process large volumes of data, develop a customized understanding of each patient's patterns, and detect anomalies in real time.

2.4 Treatments and Diagnosis

Significant variability exists in how different patients react to treatments. Genetic, biological, behavioral, and environmental factors are seen as causes of this variability, which translate into differences in clinical outcomes. Consequently, different dosing regimens may be needed for different patients. Therefore, current patient treatment regimens are becoming more personalized which contrast with the use of a "general or average" approach for recommending doses [34]. In this section, several tailored approaches for personalized treatment within the concept of smart cities are discussed.

2.4.1 Smart Medication

Technology is also changing the way that patients receive medication. There are smart medication dispensers that dispense medication when needed and blast alarm reminders, which support accurate dosing and medication adherence. Another technological feature is the emergence of 3D printed pills. 3D printed pills enable compact packaging for shipment and printing of the medication as needed, which facilitates personalized medicine and supports accurate dosing.

2.4.2 Smart Treatments

One of the major challenges in diabetes management is that patients have different responses to dose recommendations and also different disease progression characteristics. Therefore, a personalized treatment will be more effective and efficient to the patient's unique dose–effect characteristics than current trial-and-error approaches. Lee et al. [35] address the twofold dosing challenge in diabetes management by designing a novel outcome-based decision-support tool that couples a predictive treatment-effect model with a treatment planning optimization model. The strategy has shown promising results by returning better glycemic control while using less medication. In practice, the treatment reduces the risk for cesarean sections (C-sections) for moms about to deliver their babies and also the need for admission of the newborn to the neonatal unit.

2.4.3 Smart Diagnosis

The North York General Hospital (NYGH) in Canada recently implemented a real-time analytic model for biopsy decision-making for patients suffering from prostate cancer. The decision-making model allows for better decisions at the time of the biopsy and helps in preventing infections. The framework combines automated surgical systems that use big data analytic mechanisms for decision-making and an overall service system [7].

The Harvard Medical School (HMS) and Harvard Pilgrim Healthcare (HPH) recently implemented a smart diagnosis system that can distinguish between Type I and Type II diabetes. The system never uses bio samples such as blood or urine for diagnosis. Instead, the system utilizes a database with at least 4 years' worth of data from multiple sources [7].

The Mayo Clinic has implemented an advanced image processing algorithm that is capable of detecting brain aneurysms. In addition, the algorithm can measure how close is the brain from the aneurysm. The algorithm has shown to have 95% accuracy in detecting aneurysm and help to improve patient outcomes [7].

3 Personalizing Healthcare in Smart Cities: The Propeller Health Case Study

Louisville, Kentucky is one of the hardest places to live in the USA if you have a respiratory disease [36]. In the Ohio River Valley, pollution from nearby coal and oil-burning industrial facilities and car emissions accumulates quickly, reducing air quality and creating conditions that trigger asthma symptoms.

For Louisville residents like Mary Ann Stansberry, this is no small issue [37]. A few years ago, before she became a patient with Propeller, Mary Ann was visiting the emergency room at least once a year due to severe asthma attacks she could neither predict nor prevent. When she went to the ER, she was given short-term treatment, but no clear direction on how to avoid attacks in the future.

In 2015, the City of Louisville and Mayor Greg Fischer decided to embark on a project called AIR Louisville, which aimed to use digital health technology to study the impacts of pollution in Louisville and inform the city's decision-making to improve quality of life for its residents [38].

The city partnered with Propeller Health, a digital health company that develops sensors that attach to inhaled medications for respiratory diseases like asthma and COPD to track where, when, and how frequently people use their inhalers (https:// www.propellerhealth.com/). Propeller's FDA-cleared platform, shown in Fig. 2, comprises smartphone apps and web dashboards and provides actionable information on symptoms, triggers, and medication usage to improve patient self-management and inform clinical care. The platform can identify both an individual's unique environmental triggers and environmental drivers of respiratory symptoms across the community.

AIR Louisville had two main goals: One, to help local residents with respiratory disease like Mary Ann manage their disease via Propeller's digital platform and improve their daily lives by reducing their symptoms. And two, based on the millions of data points collected from participants enrolled in AIR Louisville, to give the city a data-driven roadmap to make policy changes to improve local quality of life.

The AIR Louisville program enrolled over 1200 Louisville residents with asthma onto Propeller's digital medicine platform. Residents were provided custom-built sensors to attach to their existing inhalers, which collected information on where and when they used their inhalers. The platform provided patients information on their medication use, symptoms, environmental conditions, and disease progression to their smartphones or computers (see Fig. 3). By the time the study completed, participants saw significant improvements in their clinical outcomes [38]. Participants experienced a 78% reduction in daily rescue inhaler use, an 84% reduction in nighttime inhaler use, and a 48% improvement in the frequency of symptom-free days when compared with when they started. These improvements may derive from improved self-awareness as a result of the information shared by the platform, as has been demonstrated previously [39]. AIR Louisville participants reported "increased confidence in avoiding asthma attacks" and "improved understanding of asthma" as a result of the program, such as learning about their unique environmental sensitivities [38]. Additionally, sharing data with their providers may have led to conversations about refining treatment and opportunities for early intervention.

Unlike other studies, which had generally relied on hospitalization or mortality data to understand the effects of air pollution on respiratory health, AIR Louisville was able to show people's symptoms in real time and give patients and their providers a data-driven path to personalizing their treatment plan.

AIR Louisville also revealed important insights about Louisville's air quality— among them, that many of the worst asthma "hotspots" were on the poorer, industrial west side of Louisville and that some were also downtown or on the wealthier

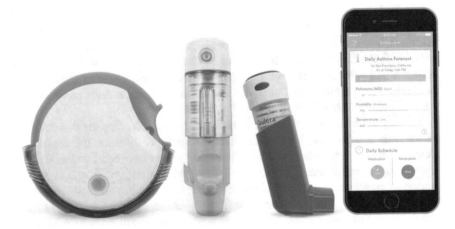

Fig. 2 Data on the timing and frequency of medication use are transmitted wirelessly via a paired smartphone or hub to Propeller's HIPAA-compliant servers. The platform, comprising smartphone apps, web dashboards, and other communication channels, presents findings and education to participants and clinicians

east side (see Fig. 3). Based on the program's findings, the researchers were able to recommend important policy changes to the city government, including enhanced tree canopy, tree removal mitigation, zoning for air pollution emission buffers, recommended truck routes, and the development of a community asthma notification system [38].

This public–private partnership served as a model for other innovative approaches to complex problems facing cities. In an interview with Politico in 2017, Louisville Mayor Greg Fischer explained his reasoning for using technology to tackle environmental issues. "Cities are platforms for innovation," he said. "What other experiments and pilots could we run around the country to increase population health? We're at a new point with these new tools we have to do this" [37].

The impact of the Propeller partnership does not stop with the city of Louisville. Armed with peer-reviewed clinical results, Propeller is now working to expand AIR Louisville through conversations with a number of other cities around the country that are interested in using a data-driven approach to address both health and environmental issues. The eventual hope is to scale the program globally, as well as to use the data to inform national environmental policy. Already, Propeller has plans to submit the results to the Environmental Protection Agency to demonstrate that air pollution standards may not be strict enough to protect the health of sensitive populations, such as asthma patients.

AIR Louisville is just one program from Propeller Health aimed at improving the lives of people living with respiratory diseases. Propeller has partnered with a California health system since 2012 on a series of studies and now a commercial program to treat patients with asthma via its digital medicine platform. In those studies, Propeller has shown that digital medicines can help increase symptom-free days [40], reduce rescue inhaler use, curb emergency room utilization, and result in high patient satisfaction [39, 41].

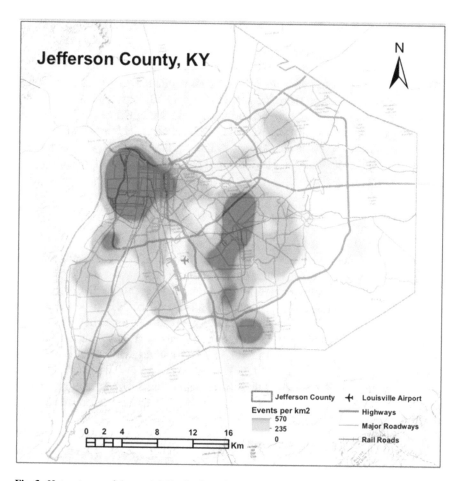

Fig. 3 Hotspot map of the spatial distribution of rescue medication events per square kilometer in Jefferson County, KY. Copyrighted and published by Project HOPE/Health Affairs as Meredith Barrett, Veronica Combs, Jason G. Su, Kelly Henderson, Michael Tuffli, and The AIR Louisville Collaborative. AIR Louisville: Addressing Asthma With Technology, Crowdsourcing, Cross-Sector Collaboration, And Policy. Health Affairs (Millwood). 2018: Vol. 37, No. 4, pp. 525–534. Exhibit 3. SABA use hotspots in Jefferson County. Source: Author's analysis. The published article is archived and available online at www.healthaffairs.org [38]

In October 2017, Propeller released Air by Propeller, an open, free API that provides local asthma conditions to anyone in the country [42]. Air by Propeller uses a machine learning model trained on millions of days of anonymized data to predict the effects of air quality on people's respiratory symptoms. Propeller Health's partnerships with Louisville, Dignity Health, and many other commercial partners are a great example of how technology can be leveraged to improve both individual and population health. With new tools at our disposal to track, measure, and intervene in the drivers of chronic disease, cities have the opportunity to invest in technology that reveals new insights on healthcare problems as it works to solve them.

4 Challenges for Personalized Healthcare within the Context of Smart Cities

The potential benefits of personalized healthcare within the context of smart cities are tremendous. However, there are also challenges and risks associated with the use of unproven treatments, services, and technology that might impact the patients and medical facilities in terms of escalating medical costs. In this section, we discuss three areas to monitor as progress is made when it comes to personalized healthcare: regulation, finances, and the development of a culture of health.

4.1 Regulation

Regulators face increasing pressure to approve the use of new personalized treatment modalities even if clinical benefits are not completely verified. For instance, in 2016 the US Food and Drug Administration (FDA) approved Exondys 51 for treating Duchenne muscular dystrophy. In this case, the FDA ruled against a scientific advisory panel and in favor of patient advocacy groups due to a lack of clinical alternatives. Many experts state that this case will set a precedent for the future approval of personalized healthcare alternatives without clear clinical benefits [43].

There are also challenges at the time of regulating clinical trials. Like any developing technology, the implementation benefits always carry some risk. However, in the healthcare setting, patients desperate for a cure are more prone to accept those risks and patient safety is often left in the hands of researchers. Therefore, proper regulation and well-controlled clinical trials are needed to keep the patients safe and to determine the efficacy of personalized healthcare.

4.2 Finances

Another issue to consider with the development of new personalized healthcare options is the cost associated with them. There is a need for finding the right balance between reimbursements and patient fees. The healthcare system will soon become bankrupt if state or private payers reimburse all interventions. Yet, if no help is provided to the patients, it will be unsustainable for patients to live longer. Policymakers must analyze existing data and work with healthcare stakeholders to find the right balance that will take care of the rapid rise in healthcare spending without removing the incentive for innovation.

4.3 Developing a Culture of Health

Patient health is greatly influenced by complex factors such as home location, and the strength of our families and communities. Therefore, personalized healthcare in smart cities should be more than the diagnosis and treatment for a disease. Personalized healthcare should be a key player in disease prevention and mitigation by facilitating ways for patients to monitor and manage their own health and also improving access to health services.

The Affordable Care Act (ACA) [44] has expanded insurance coverage in the last few years and has eliminated one of the major barriers to health care access in the USA. However, secondary barriers associated with health care access have become increasingly evident after the implementation of the ACA. For instance, transportation barriers have led to rescheduled or missed appointments, deferred care, and missed or delayed medication use. The consequences are many and some of them include the poorer management of chronic illness and thus poorer health outcomes.

Transportation barriers may prevent many older adults from obtaining needed healthcare (e.g., preventive screenings and routine interactions with healthcare providers) and exacerbate negative health outcomes associated chronic conditions. The rate of missed follow-up appointments for adult rural inpatients can be reduced by offering reliable and complimentary ride-sharing services (e.g., UBER) [45]. The availability of non-emergency medical transportation (NEMT) within a smart city framework may potentially offset negative health ramifications by affording adults with transportation to access to non-emergency medical appointments.

Also, consider the application of fall detection in a smart home. Within the smart home, ambient sensors are attached to the walls and ceiling to monitor heat-based movement with passive infrared sensors. Magnetic sensors line the doors and door frames to monitor their open and closed status. Bluetooth low energy transmitters are positioned nearby to communicate with the sensors. An elderly dementia patient is living alone in the smart home [31]. Dementia requires long-term care as patients experience disorientation and memory loss [46, 47].

The ambient sensors and machine learning algorithms detect and learn common movement patterns of the dementia patient. Research indicates that early stage dementia patients experience high variability on time spent on key activities of daily living [48]. The algorithms identified a frequent pattern of going to the kitchen at night to get a drink. The nightly kitchen visits typically last 1 min and generate 30–40 sensor events [31]. On one evening, the time spent in the kitchen becomes much longer than usual and the patient never reaches the sink. This is a deviation from the usual movement pattern for this time of day. The algorithm then identifies this as a health event which triggers an alert system to call the patient and send messaging alerts to the caregiver or an emergency care provider [49].

5 Conclusions

In this chapter, we have discussed how healthcare personalization is an opportunity for health care organizations in their pursuit of better outcomes in their service provision process. Patient heterogeneity with variability in service needs provides an opportunity to deliver more value for patients through the personalization of services. Personalized healthcare focuses on patient-centered health care, personalized health planning, and patient engagement. It encourages healthy behavior and planning. Healthcare personalization has been expanding lately due to a variety of factors including technological advances that connect data, people, and systems. Patients are given the opportunity to engage with their own health by identifying areas in their life where they are motivated to make changes for their health.

The potential benefits of personalized healthcare within the context of smart cities are tremendous. However, there are also challenges and risks associated with the use of unproven treatments, services, and technology that might affect the patients and medical facilities in terms of escalating medical costs. There is still a lot of work to do if we want the broader public to benefit from personalized medicine while minimizing both the financial and clinical risks to society and patients. Healthcare stakeholders and policymakers must continue working together to advance the personalization of medicine for the greater good of society.

References

1. Kannan P, Healey J (2011) Service customization research: a review and future directions. In: The science of service systems. Springer, Boston, pp 297–324
2. Pérez E, Ambati R, Ruiz-Torres A (2018) Maximising the number of on-time jobs on parallel servers with sequence dependent deteriorating processing times and periodic maintenance. Int J Oper Res 32:267–289
3. Ruiz-Torres A, Alomoto N, Paletta G, Pérez E (2015) Scheduling to maximise worker satisfaction and on-time orders. Int J Prod Res 53:2836–2852
4. Ruiz-Torres A, Paletta G, Perez-Roman E (2015) Maximizing the percentage of on-time jobs with sequence dependent deteriorating process times. Int J Oper Res Inform Syst 6:1–18
5. Kannan P, Proenca J (2008) Design of service systems under variability: research issues. In: Proceedings of 41st Hawaii international conference on system sciences, pp 116–116
6. Frei F (2006) Customer-introduced variability in service operations. Harv Bus Rev 84:606–625
7. Pramanik M et al (2017) Smart health: big data enabled health paradigm within smart cities. Expert Syst Appl 87:370–383
8. Kunene K, Weistroffer H (2008) An approach for predicting and describing patient outcome using multicriteria decision analysis and decision rules. Eur J Oper Res 185:984–997
9. Liu Y, Kapur K (2008) New patient-centered models of quality-of-life measures for evaluation of interventions for multi-stage diseases. IIE Trans 40:870–879
10. Shechter S, Bailey M, Schaefer A (2008) A modeling framework for replacing medical therapies. IIE Trans 40:861–869
11. Preciado-Walters F et al (2004) A coupled column generation, mixed integer approach to optimal planning of intensity modulated radiation therapy for cancer. Math Program 101:319–338

12. Pérez E et al (2010) Modeling and simulation of nuclear medicine patient service management in DEVS. Simulation 86:481–501
13. Green L, Savin S (2008) Reducing delays for medical appointments: a queueing approach. Oper Res 56:1526–1538
14. Suryadevara N, Mukhopadhyay S (2014) Determining wellness through an ambient assisted living environment. IEEE Intell Syst 29:30–37
15. HealthIT.gov (2018) What is a patient portal? Available: https://www.healthit.gov/faq/what-patient-portal
16. HIT PE (2018) Top hospital patient portal vendors by implementations. Available: https://patientengagementhit.com/news/top-hospital-patient-portal-vendors-by-implementations
17. Kruse C, Bolton K, Freriks G (2015) The effect of patient portals on quality outcomes and its implications to meaningful use: a systematic review. J Med Internet Res 17:e22
18. Cerrato P (2011) Hospital rooms get smart. Information Week Online. Available: www.informationweek.com/healthcare/clinical-information-systems/hospital-rooms-get-smart/d/d-id/1100822?
19. Alvarado M, Ntaimo L (2018) Chemotherapy appointment scheduling under uncertainty using mean-risk stochastic integer programming. Health Care Manag Sci 21:87–104
20. Alvarado M et al (2018) Modeling and simulation of oncology clinic operations in discrete event system specification. Simulation 94:105–121
21. Dzubay D, Pérez E (2016) The impact of system factors on patient perceptions of quality of care. In: Proceedings of the winter simulation conference, pp 2169–2179
22. Pérez E et al (2011) Patient and resource scheduling of multi-step medical procedures in nuclear medicine. IIE Trans Healthc Syst Eng 1:168–184
23. Sowle T et al (2014) A simulation-IP based tool for patient admission services in a multi-specialty outpatient clinic. In: Proceedings of the winter simulation conference, pp 1186–1197
24. Walker D et al (2015) Towards a simulation based methodology for scheduling patient and providers at outpatient clinics. In: Proceedings of the winter simulation conference, pp 1515–1524
25. Pérez E et al (2013) Stochastic online appointment scheduling of multi-step sequential procedures in nuclear medicine. Health Care Manag Sci 16:281–299
26. Reese H et al (2017) Improving patient waiting time at a pure walk-in clinic. In: Proceedings of the winter simulation conference, pp 2764–2773
27. Doll M et al (2015) Touchless technologies for decontamination in the hospital: a review of hydrogen peroxide and UV devices. Curr Infect Dis Rep 17:44
28. Pérez E et al (2017) Assessing catheter associated urinary tract infections prevention interventions in intensive care units: a discrete event simulation study. IISE Trans Healthc Syst Eng 7:43–52
29. Pérez E et al (2017) Catheter-associated urinary tract infections: challenges and opportunities for the application of systems engineering. Health Syst 6:68–76
30. Alvarado M et al (2017) Barriers to remote health interventions for type 2 diabetes: a systematic review and proposed classification scheme. J Med Internet Res 19:e28
31. Cook D et al (2018) Using smart city technology to make healthcare smarter. Proc IEEE 106:708–722
32. Williams J, Cook D (2017) Forecasting behavior in smart homes based on sleep and wake patterns. Technol Health Care 25:89–110
33. Kavakiotis I et al (2017) Machine learning and data mining methods in diabetes research. Comput Struct Biotechnol J 15:104–116
34. Godman B et al (2013) Personalizing health care: feasibility and future implications. BMC Med 11:179
35. Lee E et al (2018) Outcome-driven personalized treatment design for managing diabetes. Interfaces 48:422–435
36. Asthma and Allergy Foundation of America (2018) The most challenging places to live with ASTHMA. https://www.aafa.org/allergy-capitals/

37. Allen A (2017) How bourbon and big data are cleaning up Louisville. Politico. https://www.politico.com/magazine/story/2017/11/16/louisville-pollution-data-what-works-215836
38. Barrett M, Combs V, Su J, Henderson K, Tuffli M, The AIR Louisville Collaborative (2018) AIR Louisville: addressing asthma with technology, crowdsourcing, cross-sector collaboration, and policy. Health Aff (Millwood) 37(4):525–534. https://doi.org/10.1377/hlthaff.2017.1315
39. Merchant R et al (2018) Digital health intervention for asthma: patient-reported value and usability. JMIR Mhealth Uhealth 6(6):e133
40. Merchant R, Inamdar R, Quade R (2016) Effectiveness of population health management using the propeller health asthma platform: a randomized clinical trial. J Allergy Clin Immunol Pract 4(3):455–463
41. Merchant R et al (2018) Impact of a digital health intervention on asthma resource utilization. World Allergy Org J 11(1):28
42. Propeller Health (2018) Air by propeller. API. https://www.propellerhealth.com/air-by-propeller/
43. Kesselheim A, Avorn J (2016) Approving a problematic muscular dystrophy drug: implications for FDA policy. JAMA 316(22):2357–2358
44. Chen J et al (2016) Racial and ethnic disparities in health care access and utilization under the affordable care act. Med Care 54(2):140
45. Powers B, Rinefort S, Jain S (2016) Nonemergency medical transportation: delivering care in the era of Lyft and Uber. JAMA 316(9):921–922
46. Algase DL et al (2007) Mapping the maze of terms and definitions in dementia-related wandering. Aging Ment Health 11:686–698
47. Moore P et al (2013) Monitoring and detection of agitation in dementia: towards real-time and big-data solutions. In: Proceedings of the P2P, parallel, grid, cloud and internet computing (3PGCIC), pp 128–135
48. Dawadi PN, Cook DJ, Schmitter-Edgecombe M (2015). Automated cognitive health assessment from smart home-based behavior data IEEE journal of biomedical and health informatics, 20(4), 1188–1194
49. Van Sickle D, Barrett M (2018) Transforming global public health using connected medicines. Respir Drug Deliv 1:61–70

Creating an Equitable Smart City

Catherine Crago Blanton and Walt Trybula

Contents

C. Crago Blanton (✉)
Head of Strategic Initiatives and Resource Development, Housing Authority of the City
of Austin (HACA), Austin, TX, USA

W. Trybula
Trybula Foundation, Austin, TX, USA

© Springer Nature Switzerland AG 2020
S. McClellan (ed.), *Smart Cities in Application*,
https://doi.org/10.1007/978-3-030-19396-6_2

1   Introduction

The population of a city includes people with a variety of capabilities with diverse equipment available to connect with the available information sources. Some of the population will not want to participate in the effort, others will not be capable of handling the information, and others will be intermittent users of the system. Efforts in reaching *all* the people are extremely difficult as demonstrated by the recent water emergency in Austin, Texas. Heavy rains upstream from the City created a very large amount of water to move through the Austin watershed, which supplies the drinking water. The silt in the water overwhelmed the filtering capability of the water treatment plants. An emergency announcement "to boil all water before usage" was broadcast on networks, radio stations, on web sites, in text messages, and in the local papers. There were still residents that either were never aware of the emergency or chose to ignore the messages. There will always be a portion of the population that will not receive/understand/accept the messages. Technology will not solve that problem. However, there are issues and opportunities in providing for deployment of technology so that people will have the ability to receive information if they so choose. The organization (the City) needs to ensure that the capabilities are provided whether or not the individual wants to use them.

> As a human being you need two important resources if you want to survive.
> One is food. The other one is information.—Dr. Tadashi Sasaki

This chapter uses the term "smart cities" loosely, to include towns, cities, municipalities, communities, and regions that aim to use technology to create more sustainable, efficient communities that ultimately better residents' lives. In addition, this chapter assumes that smart city developments integrate digital, human, and physical systems in the built environment.

Smart approaches, defined by the British Standards Institution (BSI) as "the application of autonomous or semi-autonomous technology systems" to urban solutions are not just the implementation of networks, sensors, and devices, but the "effective integration of physical, digital and human systems in the built environment to deliver a sustainable, prosperous and *inclusive future* [emphases added] for its citizens" [1].

This chapter assumes that smart cities systems are not purely a "technical fix for an urban challenge." Human systems include dimensions such as income, literacy, socio-economic status, digital capacity, cultural, linguistic, and community preferences and norms to the practice of smart city technology design, implementation, and performance.

The physical, digital, and human systems terrain in American smart cities and communities is uneven. Digital connections between networks are unevenly distributed because digital network infrastructure lies on legacy infrastructure which often separates—or segregates—communities; sensors and Internet of Things devices rely on broadband infrastructure, commercial, public, and private broadband options, predictable service speeds and rates, and consistent access. Therefore, the axioms that govern assessment of the economic impact, return-on-investment to government, or well-being and quality of life of smart cities and communities residents can be applied only roughly to the population.

Smart cities and communities decision makers must develop competencies that consider the human dimension of cities—and the application of inclusion and equity principles ranging from compliance with Civil Rights law (or corporate diversity and inclusion standards and processes that stop short of solving for inclusion and equity as a design goal), to the ability to properly forecast technology adoption cycles based on the degree to which some communities will assimilate and adapt to new ways of interacting with the city, to the integration of diverse user systems to achieve the ideals of deliberative democracy and innovation. Underlying these competencies is that smart cities' and communities' decision makers must be able to codify, quantify, and measure the "soft variable" of trust.

The case studies presented here are based on real-world events and have been deliberated by engineers, planners, human services professionals, and city residents from every income bracket. They are meant to represent the complex and multidisciplinary environment in which smart cities and communities the technology is deployed.

It is your responsibility to develop a living point-of-view about the role of inclusion—and equity—in delivering a sustainable prosperous and inclusive future for your community. Practitioners and policy makers have the unique opportunity now to co-create the frameworks that we ultimately use to measure good—and to define the systems that can prevent dystopian realities including 'new geometries of power', political and corporate use of big data; panoptic surveillance and control of citizens and public-sector marginalization through public–private city partnerships [1]. Citizens who are engaged in smart cities processes will undoubtedly help uncover technical solutions to urban problems that respect the problems of data privacy and control, and of social inequalities and further marginalization to the benefits of smart cities that should be shared by all members of society.

> We are a community. We are all responsible for solving for equity. To solve real problems for real people, city staff, residents and technologists must collaborate.—John Speirs, Digital Inclusion Program Manager, City of Austin

2 Uneven Terrain

Smart cities technologies are being deployed on an uneven terrain, i.e., residents of smart cities are "digitally included" to different degrees. Equitable design, deployment, and assessment of smart cities technologies must consider digital inclusion. Digital Inclusion refers to the activities necessary to ensure that all individuals and communities, including the most disadvantaged, have access to and use of Information and Communication Technologies (ICTs). This includes five elements: (1) affordable, robust broadband internet service; (2) internet-enabled devices that meet the needs of the user; (3) access to digital literacy training; (4) quality technical support; and (5) applications and online content designed to enable and encourage self-sufficiency, participation, and collaboration. Digital Inclusion must evolve as technology advances. Digital Inclusion requires intentional strategies and investments to reduce and eliminate historical, institutional, and structural barriers to access and use technology [2]. Digital inclusion must also allow for the usage of older technologies. (As of 2019, the authors know of individuals still using computer technologies introduced in 2000.)

2.1   How Smart Can a City Be if Some Residents Are Not Connected?

Key elements in the smart city revolve around broadband access and the resulting digital gaps that must be closed. Despite massive amounts of public and private investment in broadband infrastructure, deep digital divides exist—rural and urban, served and underserved. The discourse related to benefits of ubiquitous broadband access in metropolitan and rural areas relies on the literature related to electrification of the grid, the Communications and Telecommunications Acts, and more recently Lifeline Services Modernization, which is the basis for the federal subsidies available to connect low-income Americans to the internet. A recently digitized set of data related to electrification of rural areas from 1930 to 1960 shows large short-run gains exceeded the historical cost of extending the grid—farmland value, agricultural productivity, and rural housing quality increased—though local wages did not [3]. The New Deal of the 1930s bore the Communications Act of 1934, which created the Federal Communications Commission (FCC). The Communications Act established the principle to make available "so far as possible, to all the people of the United States a rapid, efficient, nationwide and worldwide wire and radio communication service with adequate facilities at reasonable charges…" The prevailing belief that telecommunications channels "should serve the public and perhaps even grease the wheels of commerce" [4] helped bolster legislation that provided telecommunications infrastructure and service loans through Rural Electrification Administration. By 1996, when the Telecommunications Act established a federal Lifeline Services fund to subsidize basic telephone service for low-incomes Americans, a small percentage of the urban poor said they would go 2 days without food to keep their telephone connected. In July 2015 the FCC modernized Lifeline Services to include provision of broadband service subsidies to low-income subscribers. While there is disagreement about whether broadband can be considered a utility, there is uniform understanding in most federal and state governments that issues of economic impact and lack of access to opportunity fester when communities assume that everyone is connected [5]. And lack of information about what different agencies are doing in the smart city domain can make it harder for those agencies to coordinate sharing broadband resources when smart cities technology implementations provide them.

2.2   Why Is this Access Required?

Aspiring smart cities across the nation are grappling with increasing economic segregation, exacerbated by a lack of broadband access, a lack of sufficient speeds, and a lack of affordable options [5]. Basic needs such as access to education, access to work, and high-quality affordable healthcare increasingly require broadband access.

2.2.1   Access to Education

The "homework gap" describes the accumulation of assignments children miss because they lack access to technology or the internet. CoSN reported in 2015 that 70% of teachers assign homework that must be done online. However as late as October 2018 nearly one in five teens can't finish homework because they lack digital access, according to Pew Research [6]. Roughly one-third of households with children ages 6–17 and whose annual income falls below $30,000 a year do not have a high-speed internet connection at home, compared with just 6% of such households earning $75,000 or more a year. Twenty percent of black teens and 12% of teens overall said they rely on public access Wi-Fi to complete homework assignments. Students and parents report they take extreme measures between 6:00 p.m. and 10:00 p.m. to check messages from teachers, complete homework, or take exams. Congregating outside a community center that has Wi-Fi is effective only as long as that Wi-Fi is turned on; in cities with Wi-Fi enabled busses students can do homework in transit until the bus stops, though neither option appears safe to most. In public housing communities across the country, parents know the cost of idling in a car outside a fast food restaurant so a child can submit an assignment or parents trade services such as babysitting so that a child can use a neighbor's Wi-Fi network.

2.2.2   Access to Jobs

More than 80% of Fortune 500 companies require online job applications. This eliminates potential workers without a computer and especially without computer skills. The other limiting factor is that companies expect to have the application contain certain words relating to the job skills required. Is this a workforce digital divide point? How do small businesses handle the hiring process? There is a cost to using an external agency to screen people to fill open positions. Almost always, the small company is looking for computer skills in one skill or another.

Companies striving for a flexible workforce are increasingly in hiring "gigsters" and freelancers—individuals who are available temporarily, through electronically mediated platforms, with seemingly little risk. Some 35% of 50 million people who classify themselves as freelancers are moonlighting in addition to full-time employment. "In the sharing economy, no one's an employee, the jobs may offer flexibility and many other benefits, but traditional legal protections for workers aren't part of the package," wrote the Atlantic Magazine in 2018. Despite criticisms of the "gig economy"—especially as it relates to helping low-income people break the cycle of poverty, it remains that individuals who are not connected, who are unbanked, cannot participate as earners in the gig economy ecosystem.[1]

2.2.3 Human Services and Health Impacts of Digital Divide

By 2025, there will be more Americans over age 65 than under age 18. In 2017 the Organisation for Economic Co-operation and Development (OECD) released a study that found the gap between wealthy and low-income seniors is wider in the USA than it is in all but two of its 35 member nations—Mexico and Chile. Wealth inequality is increasing generation by generation across nearly all of the OECD member countries [7]. Low-income or semi-retired seniors who must work are similarly challenged by requirements that necessitate the connectivity required for modern communications.

2.2.4 Government

In this particular instance, government refers to the local government. Equitable and efficient governance needs to reach out to all citizens of the community. As the Internet of Things (IoT) moves from a curiosity on remembering things or answering question to providing life enhancing services, the least currently served need to be included. As health services are incorporated, communications capabilities need to be increased. A recent article in the London Times covered a proposal to move about 1/3 of its health services to internet communications via Skype as a means of relieving the current waits for medical services [8].

Built into these new services is the cost to the City to fulfill the obligations the City has to its residents. There needs to be a minimum level of service provided to all. This has two immediate implications. Service must be provided to all, which implies a very significant infrastructure upgrade. Some of the potential medical services will require high-speed connectivity that currently is only available through fiber communications. (How are older buildings retrofitted with fiber?) The second implication is that the services provided must be available equitably. How does the City handle the fact that 20% of the residents will be using 80% of the resources?

There will be a large educational effort to provide the necessary education for all residents to be able to use the tools available and to ensure that the residents can trust the connectivity is secure and their information is safeguarded. Of course, the question of "Where does the funding come from?" will be the driver behind the successful adoption.

---

[1] Gig economy workers are sometimes misclassified as contractors, fail to receive benefits and training, and may lack job security and career opportunity.

3   Changes in Work Structure: The Gig Economy

Smart cities practitioners must think carefully about the technology ecosystems—and digital demographics of workers and system users. Most workforce development pundits are bullish on the digitally powered workforce, but many legal and human resources issues remain to be worked out. Uber, Lyft, Airbnb, and other platforms that are "app economy" or "gig economy" companies contributed more than $860B to the U.S. economy in 2018 [9]. Connected Americans make extra money renting out a spare room, driving or delivering using their own cars or a company or city-owned vehicle. Thoughtful city planners will consider the degree to which their plans rely on—or could suffer from—major tenants in these new economies.

A recent study was based upon the JPMorgan Chase Institute Online Platform Economy [10] dataset to track both participation in the gig economy and earnings. They have identified around 38 million payments made across 128 distinct platforms to 2.3 million unique individuals over a 6-year period to March 2018.

They primarily analyze four key sectors of the gig economy:

- The *transportation sector*, in which drivers transport people or goods
- The *non-transport sector*, in which workers offer everything from dog walking to telemedicine
- The *selling sector*, in which people sell goods through online marketplaces
- The *leasing sector*, in which people lease out homes, parking spaces, and other assets

As this work style becomes more prevalent, the need for communications for all is important. Considering the changing lifestyle with high-rise buildings (apartments and condos) being built without parking for any vehicles. This direction for the emerging workforce style of living forces good communications. If a person is not feeling well and has no vehicle, do they call 911 (no it is not an emergency) or call Uber, Lyft, or similar company? Communication becomes critical along with the skills required to access the services.

4   The Responsibilities of the Smart City

Developing the infrastructure of the Smart City requires addressing a number of factors about the residents, the application of emerging technology, and the personnel responsible for developing and managing the infrastructure. Part of the focus needs to be on the less educated residents, which in a large number of cases is the low-income population.

4.1   Economic Mobility and Quality of Life for Low-Income Residents

There are considerations for reaching out to the low-income residents. There are two distinct groups of people that need to be addressed among these residents. There will be one group that is looking forward to an upward economic mobility and the other segment that will be complacent with remaining (or aging) in place. Each segment will require a different series of implementation steps.

Currently, there are students who do not have internet at home. Some areas provide wireless internet on school busses so the students can study on the ride to and from school. Libraries are providing internet connectivity at various locations. Malls are providing some wireless connectivity that is being employed by students. This does not address the situation where the student's family cannot afford to acquire the computer/tablet the student needs to do the work.

A number of the skills required for developing workforce capability rely on education that is available only through internet connectivity. Consequently, workforce development to provide economic mobility requires the ready access to the internet. How will the City determine the means of access? This raised the question of affordability.

The case of those citizens who are content to age-in-place is interesting in the challenges of providing the appropriate service. Cost is a strong consideration. If given the choice between internet access and either eating or keeping the home warm (in winter) or cold (in summer), the internet access cost is the limiting factor.

4.2   Identifying the Responsibilities

There are a number of agencies and people who will be responsible for various aspects of the system implementation for connectivity. Obviously, the City is responsible for the overall implementation. A recent National League of Cities report showed that 66% of cities had invested in some type of smart city technology. Program leaders often sit in a functional City department such as Planning, Information Technology, or Innovation. A key position in the City needs to be a Coordinator, which may be a dual person position. The "coordinator"—whether a job title or a department function—must understand the technologies available as well the social science required to be able to reach the least served community. Coalitions have been formed in cities like Columbus, Portland, Kansas City, and Denver to formalize and structure leadership, member rights and obligations, notions of reasonable transparency, and how to operate in the public realm. Obviously, a national organization needs to drive a set of standards for implementation that provides both the connectivity capability required as well as the security level (trust) needed to relieve concerns of the users. Often forgotten, the commonality that standards for the usage of applications increases the diversity of skills required for the users.

4.3 Who Provides the Funding?

There is tension over who should pay to connect underserved Americans. What is the right proportion of public vs. private investment? At a recent congressional hearing, the American Cable Association said more than 100 million homes have access to broadband speeds over 100 mbps and only 5.3 million remain with speeds less than 5 mbps. Many cable providers claim to have provided cable to 800,000 homes that would normally be eligible for FCC Lifeline Services support. Some providers push for federal funding to connect unserved high-cost areas, while many government-funded physical infrastructure projects, such as bridges and roads, lack broadband connectivity which could make them safer, smarter, and extend their useful life.

Since the passage of the 1996 Telecom Act, broadband companies have invested roughly $1.6 trillion in wireline, wireless, and other broadband technologies. But closing the divides among and within cities requires investment that many Internet Service Providers (ISPs) are not often willing to make. Public–private partnerships, such as one in San Francisco, will ensure that 150,000 people who cannot normally afford it will have high-speed broadband. Municipal broadband networks are provided fully or partially by local governments, some exclusively for municipal services and others for public access. The Community Broadband Networks Initiative of the Institute for Local Self Reliance reports that 800 U.S. communities have some form of municipal network or cooperative. A recent Harvard study showed that municipal broadband service cost is 50% lower than commercial broadband [11].

The implementation of fiber is not quick or straightforward. There are areas of Austin that had fiber installed run in the streets in early Fall 2017 with a promise of service by the end of that year. As of the writing of this chapter (Spring 2019), connectivity to the homes is spotty with a system's issue preventing completing the coverage. The coverage of fiber showed a distinct economic divide in Kansas City, where the initial fiber installation was on one side of Troost Avenue (moderate incomes) and not on the other (poorer families). These inequalities must be considered and addressed if there is to be complete coverage.

One area of concern is that internet coverage maps may be inaccurate. Areas in Austin, Texas, show internet connectivity available in some neighborhoods, but the high-speed access is not really present. If the future requires the capabilities only available through fiber, how do existing apartment buildings address this need? Existing Cat 6 cable might be installed, but is that capable of handling the medical interactions over the internet? How can a fiber cable be retrofitted in existing buildings? If the answer is that it is not practical, then there will be a group of residents who will not have the full service.

5   Issues with Data

The concept of information is about both what the information (data) contains, who owns it, and how the data is communicated. The content of the information, the data, is dependent on which organization is collecting it, what business they are in, and the level of oversight on critical and personal data.

5.1   *What Is Connectivity?*

Congress is struggling to determine how to allocate resources to the underserved because there is no single map of broadband access, penetration, and speed in the USA. The U.S. National Broadband plan tracks resident access to service in cities across the USA, as reported by the ISPs themselves. All facilities-based broadband providers are required to file data with the FCC twice a year (Form 477) on where they offer Internet access service at speeds exceeding 200 kbps in at least one direction [12]. Connected Nation was founded in 2011 and began its own study of internet access and speed—including upload and download speed—and found glaring differences between what the Internet Service Providers (ISPs) were reporting and what was available in communities around the country. There is no federal agency responsible for monitoring or regulating ISP statements about service levels provided to residents in various zip codes.

The bipartisan Congressional Smart Cities Caucus in 2019 is sponsoring legislation to address the ten million Americans in urban areas that do not have access to broadband.

In 2017, the National Digital Inclusion Alliance, a non-profit that serves as a "unified voice for digital inclusion access," including broadband access, digital literacy, and devices, published a groundbreaking study which has implications for other cities [13].

Based on a mapping analysis of FCC Form 477 block data and city permit records, AT&T's Digital Redlining of Cleveland showed that AT&T had quietly ignored the service areas of four inner-city wire centers when it installed its high-speed "fiber to the node" VDSL network throughout most Cleveland suburbs and better-off city neighborhoods. The now-standard VDSL infrastructure, which combines optical fiber running to neighborhood locations with upgraded copper lines the rest of the way to customer homes, provides most AT&T households with Internet access at download speeds above 24 mbps, as well as the option of IP video "cable" subscription service. But residents of neighborhoods served by the four inner-city wire centers—Hough, Glenville, Central, Fairfax, South Collinwood, St. Clair-Superior, Detroit–Shoreway, Stockyards, and other high-poverty areas—are still relegated to AT&T's old, much slower copper-only ADSL network.

The result, according to AT&T's own 2016 data reported to the FCC: 55% of Census blocks in the city of Cleveland had maximum AT&T download speeds of 6 megabits per second (mbps) or less, and 22% had download speeds of 3 mbps, 1.5 mbps, or 768 kbps.

NDIA and CYC pointed out that AT&T was free to discriminate against these neighborhoods, and keep the practice secret, because of its success in lobbying Ohio legislators to eliminate municipal cable franchising and oversight in 2007.

In 2007, AT&T succeeded in lobbying the Ohio General Assembly to eliminate municipal franchising of cable television providers by dangling the promise of a new era of "cable competition" in communities throughout its service territory... AT&T's "cable franchise reform" legislation explicitly permitted providers under the new state-run video service authorization system to serve less than 100% of their designated service territories—a provision that led critics like the City of Cleveland to warn of the exclusion of poorer neighborhoods... AT&T dismissed the idea that providers would redline or cherrypick communities, and legislators apparently believed them; the legislation passed both houses with virtually unanimous support, including "Yes" votes from every Cleveland representative.

AT&T's Digital Redlining of Cleveland got noticed. In the weeks following its release, the report was widely covered and linked on news and tech industry sites including The Hill, BuzzFeed, Grio, Ars Technica, Engadget, Gizmodo, Fierce Telecom, and MultiChannel News among others. Cleveland alternative newspaper Scene made it the subject of two long, detailed articles, and the city's local ABC news outlet, WEWS, covered it as well.

An NDIA affiliate in Dayton, Advocates for Basic Legal Equality, requested a similar map of AT&T's deployment pattern in the Dayton area and released it publicly on March 22, with the headline "AT&T Fails To Invest in Low-Income Montgomery County Neighborhoods."

On March 15, new FCC Chairman Ajit Pai told an audience in Pittsburgh: "Just last week, a study of broadband deployment in Cleveland suggested that fiber was much less likely to be deployed in the low-income neighborhoods. This highlights the need to establish Gigabit Opportunity Zones..."

In August, an attorney representing three Cleveland AT&T customers submitted a formal complaint to the FCC citing the report and alleging that "AT&T's offerings of high speed broadband service violates the Communications Act's prohibition against unjust and unreasonable discrimination."

In early September, NDIA released new maps showing very similar patterns of broadband deployment discrimination by AT&T against high-poverty neighborhoods in Detroit and Toledo. Two AT&T customers in Detroit soon joined the attorney's FCC complaint. (The customer complaints are currently the subject of a confidential mediation process between the attorney, Daryl Parks, and AT&T.)

In late October, the FCC released its Notice of Proposed Rulemaking on the Lifeline program, including a section on "Digital Redlining." "Recent reports argue that some service providers engage in 'digital redlining' in low-income areas—a practice that results in certain low-income areas experiencing less facilities deployment when compared to other areas, and that low-income consumers in those areas may experience increased difficulty obtaining affordable, robust communications services… We seek comment on how the Commission can address this issue with the Lifeline program." (Here are NDIA's comments on this section, among others.)

Through all this, AT&T's Digital Redlining of Cleveland has continued to be discussed and linked by the media, most recently in this Fast Company article the day before yesterday. And through all this, AT&T spokespeople have continued to respond with non-denial denials like this and this:

"We do not redline," AT&T regulatory and state external affairs executive VP Joan Marsh reiterated Wednesday following the complaint's filing. "Our commitment to diversity and inclusion is unparalleled. Our investment decisions are based on the cost of deployment and demand for our services and are of course fully compliant with the requirements of the Communications Act."

There is no single definition of broadband and official plans may refer to a variety of criteria. Wikipedia's article, National Broadband Plan, says, "Broadband is a term normally considered to be synonymous with a high-speed connection to the internet. Suitability for certain applications, or technically a certain quality of service, is often assumed. For instance, low round trip delay (or 'latency' in milliseconds) would normally be assumed to be well under 150 ms and suitable for Voice over IP, online gaming, financial trading especially arbitrage, virtual private networks and other latency-sensitive applications" [14].

Broadband is "technology-neutral." While there are differences in performance and capabilities, broadband can be delivered by Digital Subscriber Line (DSL), Cable Modem, Ethernet, Fiber, Wireless, Satellite, or next generation technologies.

By 2012 more than 20 countries were using national broadband plans to evaluate comparative and competitive advantages of their national broadband. And today all G7 countries have a *national broadband plan* in place or under development.[2] *Connecting America: The National Broadband Plan* was established under the American Recovery and Reinvestment Act of 2009. It directed the FCC create the plan and to "include a detailed strategy for achieving affordability and maximizing use of broadband to advance consumer welfare, civic participation, public safety and homeland security, community development, health care delivery, energy independence and efficiency, education, employee training, private sector investment, entrepreneurial activity, job creation and economic growth, and other national purposes."

A 2010 U.S. Government Accountability Office (GAO) report ranked the USA 15th of the 30 OECD countries for both broadband deployment and adoption, down from the top third in 2007. The GAO report determined that some of the U.S. top trading partners such as Canada, France, Japan, and South Korea had used public–private partnerships, competitive market policy, consumer subsidies, and digital literacy training to increase competitiveness and utilization of online government services. The OECD's Broadband Portal provides policy makers with a range of

---

[2] The Canadian Federal government and territories agreed to develop a national plan in October 2018.

indicators which reflect the status of individual broadband markets, including coverage, penetration, and speed, but also usage and prices and mobile termination rates. The World Bank Broadband Strategy Toolkit assists in policy development in developing countries. Furthermore, the USA and Europe closely align wired broadband planning with smart grid planning for enablement of energy demand management and national security.

The 2010 U.S. Broadband Plan recommends that the USA adopt and track six goals to "serve as a compass," for completion by 2020. The goals are in short:

- The most effective and efficient ways to ensure broadband access for all Americans;
- 100 million U.S. homes should have affordable 100 × 50 mbps service;
- The USA should lead the world in mobile innovation, by coverage and speed;
- Every American should have affordable access to robust broadband service, and the means and skill to subscribe *if they so choose* [emphases added];
- Every American community should have affordable 1Gbps service to an anchor institution such as a school, hospital, or government building;
- To ensure the safety of the American people, every first responder should have access to a nationwide, wireless, interoperable broadband public safety network;
- Every American should be able to use broadband to track and manage energy consumption.

By 2019 the National Telecommunications and Information Administration (NTIA) announced that more than 20 federal agencies had set out strategies to streamline permitting, developed comprehensive and shared broadband availability data and evaluate how to more effectively allocate government-owned spectrum. NTIA's BroadbandUSA program serves communities, industry, and non-profits that want to expand broadband infrastructure and promote digital inclusion.

5.2   Whose Data?

There is an ongoing issue across the globe on who has rights to the data being collected. On one side, the major data accumulators will respond with the fact that they atomize the data so that no one can be identified for the "cleansed identification characteristics." There are researchers who claim that re-anonymizing data to identify individuals is not difficult [15]. The other side of the discussion is that all the information about me as an individual should belong to me and require permission for other people to use it. Europe has been leading the legislative way in this particular issue.

6 Critical Decisions

Dr. Tadashi Sasaki, who famously developed[3] the liquid crystal display, worked on vacuum tubes, and developed the microprocessor, invoked the human dimension of technical work. In 2008 he keynoted a nano-technology convention in Dallas. I interviewed his deputy, who had been designated to speak for him. Japan and the USA had been in an intense race to dominate the global semiconductor industry, but increasingly without global research and development joint ventures little progress toward developing next generation computer chips could be made. When technology is very new, I (primary author) asked, how do you know a joint venture will be successful? "When your partner's aim is to improve the human condition, the technology will be successful." The basis for technology adoption? Trust. "Every morning I go to the barber. I lay in a chair. The barber puts a towel over my eyes. He puts a razor to my throat. Why am I not afraid? Trust!" The people, he said, must trust technology companies as much as he trusts his barber.

At the core of the smart city is the notion that information can be used to better the lives of its residents. A smart city can be conceived as the goal for urban development—'a good place to live,' 'a healthy city,' and primarily as a way to support policies related to urban development [16]. Another narrative focuses on smart cities' ability to compete against other cities [17]—to attract investors, help the corporate and small business sector compete more effectively, sustain the environment, operate more cost-effectively.[4] Another branch of literature on smart cities focuses on governance. A smart city is a "two-way conversation," enabling the City and residents to share information, make decisions about how to allocate resources, how to adjust to the changing environment, how to collaborate to make important decisions, and how to innovate together.

The competent smart city planner makes a case for inclusion and equity in user requirements materials, procurement documents, and other artifacts of the engineering process; the trustworthy smart city engineer has thought carefully about how values like equality and fairness hold up in the product and the technology roadmap. A capable smart cities team has a shared and thorough understanding of the substrate upon which smart city technologies are deployed—the community's digital capacity: broadband access, digital literacy (e-readiness), and the cocktail of device-operating or system stack layers in the hands of city workers and residents.

---

[3] Sasaki said he respected Bob Noyce because Noyce talked about his success in the "development" not "the invention" of the microprocessor. Sasaki said he didn't invent the microprocessor idea. A Nara Women's College software engineering researcher made the original design proposal in a brainstorming session; his team and ultimately he rejected the idea. He lost track of the woman who had the original microprocessor design idea.

[4] Through the use of big data the city can be controlled in real time, for example. This literature often fails to acknowledge or address elements of culture, politics and policy.

6.1 Smart City Decision Makers Solve for Inclusion and Equity

What the decision makers need to understand is that the diversity and inclusion are not fixed perspectives, and there is no standard. This leads to differing views about the roles of inclusion and equity within the community's perspective. It might be advantageous for the decision makers to start out examining what is meant by "inclusion" and "equity" in their specific City—and from whose perspective?

6.2 Three Kinds of Decision-Making for Inclusion and Equity

Many of the top smart cities and communities projects are in the five U.S. states whose populations are changing most quickly—Texas, California, Florida, New York, and Illinois. These states are also the states with the earliest (save New Mexico) minority–majority urban areas.

In the past decades, there has been considerable effort to determine whether diversity and inclusion policies are good for business. The corporate world has taken a variety of approaches to determine if there is a relationship between diversity and inclusion and company performance, by evaluating individual performance, business process performance, and shareholder value. Some studies focus on improved employee recruiting and retention as a measure of success; proponents of supplier diversity policies and programs point to increased supply chain stability and improved ability to source and integrate supplier innovation. Since 2007, studies focused on patenting show that mixed-gender patent teams outperform single-gender patent teams in patent production, citation, and revenues. (Single-gender male and single-gender female teams perform about the same.) Catalyst and others analyze shareholder value and market performance, finding that Boards of Directors with more than three women outperform other boards financially—median gains in Return on Equity (ROE) of 10% points and Earnings Per Share (EPS) of 37%. At the same time, most Silicon Valley tech giants and start-ups struggle with diversity and inclusion.

Experienced and emerging leaders are constrained by their lack of decision-making experience using a diversity and inclusion lens. Most fail to build expertise because they often lack an in-depth understanding of their personal and professional perspectives on the implications of diversity and inclusion practices. Having a personal perspective on diversity often starts with an awareness of oneself. A library of research shows that though we aspire to be objective, our decisions are prone to bias [23]. The highest performing leaders—whether or not their perspectives agree—do have a v the impact of demography and cultures on business performance. High-performing technology leaders don't follow the "Hollywood script" when it comes to evaluating equity for the projects they serve. They develop a personal and professional perspective on diversity and inclusion issues, policies, and practices as it applies to their work. They have systematic and methodical analytical approaches for weighing the causes and costs of ignoring diversity and inclusion perspectives when they make business decisions. As members of multi-dimensional teams solving layered technical problems, they become good at asking questions about how smart city solutions solve for equity.

"Why should companies concern themselves with diversity?" ask Thomas and Ely in their 1996 Harvard Business Review article, Making Differences Matter: A New Paradigm for Managing Diversity. Demographic differences in and of themselves don't deliver on the legal, moral, and economic promises of diversity, "how a company defines diversity—and what it does with the experiences of being a diverse organization—that delivers on the promise." Thomas and Ely studied corporate diversity programs over 6 years and describe three diversity and inclusion perspectives that unite companies, irrespective of industry. These paradigms can be considered roughly to relate to smart city and community motivations and capacity with respect to the human dimension of smart city technology and the values placed by smart city communities on inclusion and equity.

The archetypical decision makers presented here are symbolic, though readers in different industries, countries, or with different overarching objectives may recognize the motivations, thought patterns, preferences, and aspirational goals. Each archetype also embodies an era of diversity and inclusion practice, legal environment, and organizational performance orientation.

These archetypes should not be used to stereotype or to develop a list of "ten things to never or always do"—archetypes embody patterns in a larger narrative. Get to know these archetypes to develop your critical thinking skills so that you can figure out which kinds of problems and solutions decision makers will prefer and how they will communicate about them. Then you'll be able to apply your own perspective to the smart cities projects you are working on.

- "We're on the same team. We are colorblind. You will be treated fairly."—The Guardian
- "You can be yourself here and still succeed. We see things from your point of view."—The Advocate
- "We innovate because we are different. Bring your differences to your work and the decisions you make so that we can learn and innovate."—The Innovator

The Guardian decision maker is focused on anti-discrimination, fairness, and compliance with federal law such as the Civil Rights Act (1964), the Equal Employment Opportunity Commission, and the Americans with Disabilities Act. It is not well known that at the beginning of World War II, the National Urban League President Lester Blackwell was charged with integrating black and white unions, especially those that contracted with the federal government. At the tail end of World War II, officers' quarters had begun to integrate, because it was not cost effective to build separate quarters. The segregated military units, such as the Fighting Irish, for example, were disbanded. Executive Orders were issued starting in 1948 that required the Department of Defense to buy from diverse suppliers. In 1964, Whitney Young, the most prolific leader of the National Urban League called on then President Nixon to "create an economic bill of rights." Nixon was not able to recruit Young to lead a new "black capitalism" initiative at the Small Business Administration. The Guardian aims to root out discrimination in formal processes, for example, by ensuring that diverse candidates can apply for jobs without barriers, often by ensuring that hiring managers and staff are "color blind." The Guardian believes that marginalized individuals, if given the chance to "get in the front door" will assimilate to the company culture, and that assimilation as a process is something that is a burden equally shared by all employees. To help employees assimilate, the Guardian may endorse mentoring programs and training that ensures all employees treat each other fairly and according to the law. The Guardian may focus on numbers, such as the number of complaints, lawsuits, or reporting criteria expected by the EEOC. The Guardian may demand that technologies procured by the organization are compliant with the American Disabilities Act Section 508, which requires Federal agencies to ensure electronic and information technology is accessible to people with disabilities including employees and members of the public.

The Advocate decision maker celebrates and advocates for differences, believing that diverse populations should have access to opportunity and the company should legitimize diverse population groups. The Advocate measures diversity and inclusion program success by how well the company recruits and retains employees, and how well the company leverages differences to gain insight and access to non-traditional (diverse) suppliers and expanded access to diverse markets. The Advocate is likely to embrace employee satisfaction survey questions such as "I can be myself and still succeed," and "my manager values and respects differences in employees," and training that focuses on sensitivity to and acceptance of differences while not necessarily leveraging those differences for an improved bottom line. The Advocate may support the use of Employee Resource Groups or the so-called Affinity Groups as a way for employees to feel a part of their community while being a part of the enterprise of the whole. Frito Lay extolls the value of diversity in innovation—its best-selling brand of Dorito's was developed by food scientists who interviewed Dorito's delivery drivers about various flavors. The Advocate may endorse "reverse mentoring" in which a junior employee from an underrepresented group mentors an executive.

The Innovator decision maker sees diversity and inclusion as a means to enabling a first-class learning organization, for both organizational effectiveness and innovation. The innovator believes that different perspectives and approaches to work are valuable; diverse perspectives should be harnessed to improve the business. Training focuses on recognizing that employees bring their cultural backgrounds to the decisions they make; employees are encouraged to see diverse perspectives as a learning experience. A 2010 AT&T recruiting advertisement shows a group of professionals of different genders and clearly of different ethnic backgrounds, "It's not about counting people," the advertisement says, "It's about making people count." The innovator may focus on measuring the success of diversity and inclusion practices by evaluating managers' cultural competence, cultural excellence, and the degree to which teams are inclusive.

Many U.S. and global technology companies have experimented with assembling engineering and innovation teams based on superficial, visible characteristics of diverse teams—race, ethnicity, gender, age—and less visible characteristics—socio-economic background, religious background, diversity of family size, and even orientation to time, i.e., linear vs. cyclical and short vs. long-term orientation.

Patents are a good way to capture ideas, whether they are improvement patents or an entirely new patent. In the creativity literature IQ is not a good predictor of inventiveness because it measures convergent thinking.[5] Convergent thinking relies on an individual's ability to answer logical questions based on a finite number of options. Divergent thinking—the ability to think laterally, to identify unexpected relationships and ideas in response to a question—to analogize and to "think outside the box" is most associated with creativity and innovation.

Early studies on differences and innovation focused on gender and race. For example, a 2007 report called "Who Makes IT?" looked at gender patterns in information technology patents. It turns out that 80% of U.S. IT patents are made by U.S. or Japanese inventors, so the study was restricted to those countries. In 2007, 9% of U.S. patents had just one female inventor. But mixed-gender teams have citation rates 26–45% higher than single-gender male or single-gender female teams. Whether the company had relatively few female inventors (5%) or a relatively large percentage (25%), male only and female only patent teams perform about the same. Subsequent studies have shown that experience and multiple knowledge domains are most responsible for high-performing innovative teams.[6] The more diverse the knowledge and input that are applied, the more novel the output. Knowledge that is too focused, applied deeply to a domain in which the team members have expertise can lead to predictable solutions, but can also result in "competence traps."

How do these distinct perspectives show up in discussions about designing, developing, and deploying smart cities technologies? In what cases is one perspective appropriate and in which cases could all three perspectives be valued? What steps must be taken to operationalize these perspectives? Who has the authority to say that the community has met the standard for considering the dimensions of inclusion and equity in smart city work?

6.3   Broadband: A Lot of a Little or a Little of a Lot?

Too often communities face tough choices about how to allocate broadband resources equitably. Internet service providers and eligible telecommunications carriers serve low-income communities for various reasons, ranging from corporate philanthropy to FCC regulation.

---

[5] Intelligence Quotient (IQ) measures an individual's ability to learn and use existing knowledge. Empirical studies show a low correlation between intelligence and creativity. Modern creativity research increasingly focuses on defining attributes of creativity—apart from IQ—and their relationship to innovation and economic performance.

[6] Superman or the Fantastic Four? Knowledge Combination and Experience in Innovative Teams, Taylor and Greve, 2006.

Thinking of your smart city project, how would you solve the following case?

You and your team have been working for the last year to identify broadband providers that can serve 1000 very low-income households in your district. Two large internet service providers have flown in to make special offers.

- Sally from Big Service Solutions says, "The best way to serve your community is to provide the best bandwidth to a few households. We'll give you 1000 mbps symmetrical connections for 100 households at no charge."
- Bob from Jump Cable Co says, "We can do better than that! We'll provide 5 mbps symmetrical connections to each and every one of your 1000 households."

Both providers want exclusive rights to provide internet to the community; Big Service Solutions will sign up residents who don't have internet for 1000 mbps service packages offered at $150 per month, about 10–15% of the average household income; Jump Cable Co will offer upgrades to residents, from the 5 mbps symmetrical service to a 30 mbps service for $80 per month.

The team must determine whether to take Sally's or Bob's offer.

- "Take the lower bandwidth in every home. We need no more than 5Mpbs to enable communication with many city services, especially our new demand-response program, which will enable the City to save money by adjusting each resident's thermostat when the price of electricity soars. Also, ensuring that every home has enough broadband for a child to do their homework at home will undoubtedly create a strong workforce."—City utility manager
- "Only 50% of our households are even interested in being connected to the internet and 1000 Mpbs will go to waste for people who only know how to check email and Facebook. Give the very high-speed broadband to homes that can use it. A very high-speed connection will help our residents with chronic illness and disabilities participate in telemedicine, reducing the number of trips our ambulance drivers have to make. And tech-savvy small business owners will use their 1000 mbps connection to sell all over the world—what a boon for economic development! The well-connected will inspire their neighbors to invest in low-speed broadband, will help neighbors who can't."—Fortune 500 technical director

Naturally the discussion turns to an assessment of whether the city could build and manage its own network, how to provide free internet access inside community centers or in limited or ubiquitous public areas, and naturally which wireless carriers have the best discounts. Even in the best-connected cities, smart city leaders will likely be faced with decisions about how to distribute broadband equitably.

6.4 Sensors and IoT: Can Technology Be Racist?

When I (primary author) teach as a part of interdisciplinary graduate program focused on the Human Dimensions of Organizations, the case method forces students to make tough decisions from diverse perspectives. In the "For All Texans" case study, students assume the perspective of a team working at a state agency which has procured kiosks to enable Texas residents to comply with small business regulations more timely and cost-effectively. Consumers will be able to apply for state sales tax, file and pay for permits, and remotely apply for licenses with the assistance of a video camera. "There's just one problem," the case study reveals, "the kiosks don't seem to be working for African American consumers." The team is directed to a YouTube video that has three million views as of this writing [18]. "Black Desi," and his co-worker, "White Wanda," use the Hewlett-Packard MediaSmart computer, a "state of the art computer." Desi says the face-tracking software should follow him as he moves. It does not. "My white co-worker Wanda is about to slide into the frame. As you can see, the camera is panning to follow Wanda's face... As soon as black Desi gets in the frame? Nope. No face recognition anymore, buddy."

The team is asked what could explain this situation, who is to blame, whether technology can be racist, and how the problem should be resolved. Graduate students with similar technical backgrounds and tenure still offer varying approaches to solving the case study:

- "The truth is it's a technical problem. The technology is built on standard algorithms optimized for 18% grayscale. There may have been insufficient foreground lighting.[7] Technology is inanimate—it can't be racist. The state agency is not responsible, the vendor is responsible. The State's role is to procure the best technology at the best price, not to tell manufacturers how to design their products. We'll have to find alternatives so to maintain our mission to serve all Texans."—Oil and gas executive
- "The technology is racist because it discriminates against people with darker skin tones; if diverse engineers had designed, tested and deployed the product, the firm would have discovered the issue before the kiosk was in the field. The specification we issued was clear that the tool would be used by all residents of the State. We should put a hold on this program until these issues can be worked out."—Military human resources leader

Case study participants learn to dynamically shift perspectives so that they can choose the best "leadership perspective" to respond to the case. Is there a comfortable ending for this discussion? Would the Guardian, the Advocate, and the Innovator agree on what a comfortable ending might be? HP was lauded for its response posted on a company blog within a week: "Everything we do is focused on ensuring that we provide a high-quality experience for all our customers, who are ethnically diverse and live and work around the world. We are working with our partners to learn more." The post linked to instructions on adjusting the camera settings, something both Consumer Reports and Laptop Magazine tested successfully in Web

---

[7] "The technology we use is built on standard algorithms that measure the difference in intensity of contrast between the eyes and the upper cheek and nose. We believe that the camera might have difficulty 'seeing' contrast in conditions where there is insufficient foreground lighting," Hewlett-Packard said.

videos they put online [19]. But HP has not been alone. In 2010, Taiwanese-American Joz Wang discovered her Nikon Coolpix360S had a "bias for Caucasian faces," because she received an error message, "Did someone blink?" when her picture was taken. Nikon also posted a fix quickly, saying that the camera software had been optimized for the U.S. market. HP, Nikon, and Sony did not provide insight into the glitches in the weeks after product release [20].

While facial recognition software and tools have improved dramatically since 2010 [21], it's important to note that there remains a wide spread of capability across the industry. Issues related to hardware, software, algorithms, markets, and product development management processes are all at play. Smart city planners and technologists can't assume that large companies, even those that are deeply committed to equity and inclusion, will catch these kinds of unintended outcomes.

7 Unlocking the Connection: The Housing Authority of the City of Austin

In 2014, Google Fiber's anticipated announcement was revealed. The second U.S. city selected by Google Fiber to provide service would be Austin, Texas. Google Fiber also said it would provide 100 Community Connections—1Gbps symmetrical internet service connections—free for 10 years. The City of Austin's Digital Inclusion Department pulled together a team to evaluate more than 4000 non-profit and anchor institution applications. The Housing Authority of the City of Austin was awarded a Community Connection for the community center of its largest family property—216 households at Booker T. Washington Apartments.

The Housing Authority of the City of Austin (HACA), established under the 1937 U.S. Housing Act, which then—U.S. Congressman Lyndon Baines Johnson was instrumental in passing, is home to about 5000 people in public and subsidized housing and 14,000 Housing Choice Voucher (Sect. 8) holders. Residents in public housing earn on average $10,000 to $14,000 per year. Sylvia Blanco, HACA's Executive Vice President began to wonder why every resident in public housing couldn't have access to the internet. A 2014 Choice Community Neighborhoods survey showed that at two properties, about 10% of residents had an internet connection—a stark contrast to other residents of Austin, which in 2007 was named the most connected city in the USA, and in 2014 showed that 92% of Austin residents had an internet connection.

Google Fiber agreed to provide a free internet connection to every household in its service area; funding from Ford Foundation and Open Societies Foundation enabled the program to secure a program lead, as well as funding for training by Austin's public access computing agency Austin Free-Net and evaluation by the University of Texas at Austin's Moody College of Communication. Google Fiber's focus on the three legs of the stool: a connection, digital literacy, and a device helped ground the effort.

On November 21, 2014, U.S. Housing and Urban Development Secretary Julian Castro visited Austin to help HACA launch of a first-of-its-kind initiative to connect every single housing authority resident to the internet, digital literacy training, and computers which would be earned by residents who participated in digital literacy training. (His visit inspired the White House-HUD initiative ConnectHome.)

Numerous partners provided additional funding and resources. Notably, Austin Community College began to donate computers which would otherwise have been recycled. A local technologist developed a way to reimage those computers using Clonezilla and Linux with LibreOffice, an open source software that had antivirus protection than any commercial software provider at the time. (The computers being donated had so little RAM, no current operating system would run on them.) To guard against the possibility that some residents who earned devices taking classes would not have internet at home, World Possible's suite of offline interactive content, including Wikipedia, the Khan Academy, Great Books of the World, MIT's Scratch, and a variety of games and interactive learning tools, were installed on the computers.

The first phase of the program focused on three public housing properties in Google Fiber's first service area. Residents were required to register directly with Google Fiber. Paper flyers and applications were the norm in public housing. Many residents did not have an email address, a requirement to set up an account for the service. Those who did have an email address often did not have experience with two-step authentication, required for almost any account registration, app installation, and online banking services. With the support of EveryoneOn, HACA staff organized Tech Ferias (Tech Fairs) to inspire residents to take advantage of online services and the possibilities of the internet. Austin's Independent School District attended to teach parents how to access their children's records through the AISD web portal. CapMetro, the local public transportation authority, demonstrated the app that residents could use to plan their trip and locate their bus.

Many residents still had concerns. "What if I make that doctor's appointment online? When I go to the doctor and they say I don't have an appointment, I'll have no proof." "My bank teller knows me. If I make a deposit online and it disappears, how will I get it resolved?" Some worried about the internet disintermediating their relationships with family and friends. One elderly woman with a disability said that if she were able to pay her light bill online, her sister wouldn't visit to take her to the payment location.

HACA staff invited Google Fiber and Austin Free-Net staff to canvas together. Three staff—each from a different organization—at one resident's door may seem like an inefficient use of resources. But residents needed assurance that yes, your internet connection is really free, and you can learn how to use a computer and yes, when you get that receipt from your online payment you will store it on your own computer which will stay in your home. Trust is the key to technology adoption.

Residents who had previously shied away from using the internet registered. Several reported that they didn't use the internet often, but that it drew their families closer. One grandmother reported that she got to see her grandchildren more often as her daughter, a local college student, skipped the library computer lab and completed her homework at her mother's home while her children watched television.

Digital literacy classes included residents with a wide variety of skills and experiences. Residents who had had bad experiences in school or who said they couldn't learn either eagerly or reluctantly enrolled in a 60-h training course called "Tech Starters I," while other residents who enrolled were self-taught programmers, had built websites or computers as a hobby, or used complex software in their work environments. Austin Free-Net's approach was to teach whatever a student wanted to learn first—whether or not the student had the pre-requisite skills to learn it. Classes were loud, noisy, fun-filled, serious, and collaborative.

In every class, we began to look for natural social problem solvers. The resident who looks around to see who needs help, who translates for a neighbor, who provides the missing context about why this resident wants to use that website. These residents were invited to become Digital Ambassadors. They would earn a stipend to volunteer to help in the classroom, to promote internet adoption and digital literacy, and to work on special projects, such as developing a class on parental controls, creating a Google calendar showing when food banks are open, or teaching Linux to neighbors on the weekend. People transfer technology. Among HACA's Digital Ambassadors are those that excel in working with various segments of residents who adopt technology and innovations at different rates: the Innovators (Techies), the Early Adopters (Visionaries), the Early Majority (Pragmatists), the Late Majority (Conservatives), and the Laggards (Skeptics).[8] The City of Austin's Digital Inclusion Grants for Technology Opportunities Program (GTOPS) enabled HACA resident [Computer] Lab Apprentices assisted hundreds of residents in their community center computer labs.

The HACA Scholarship Foundation had awarded almost one million dollars in scholarships by 2016, but in 2017 fewer than half of scholarship awardees striving to break the cycle of poverty through education had a device at home. The City of Austin's PC Community Loan program pilot provided loaned refurbished laptops to scholarship awardees, and desktops to two other groups. In addition to providing basic access to internet, devices and digital literacy in the first phase, HACA was fortunate to secure a full-time Digital Inclusion Fellow through the Nonprofit Technology Empowerment Network (NTEN). NTEN's Fellows, posted in seven cities around the USA received special training to design their own program. Through this program HACA was able to provide special digital inclusion support for 130 participants of a HUD-funded Family Self-Sufficiency program, providing desktops, webcams, and training so that program participants could interact with case workers using video conference calling and screen sharing. In its first 2 years a Work Study Internship program with Austin Community College provided HACA almost 20,000 h of IT support to refurbish, reimage, and deploy donated computers; IT Associate's degree candidates are "near peer" mentors for residents.

---

[8]The technology adoption lifecycle describes adoption or acceptance of new ideas, technology or innovation. In 1957 Beal and Bohlen developed the model for the Study of the Diffusion of Farm Practices. Everett Rogers describes how and why innovations spread in his 1962 book Diffusion of Innovations. Geoffrey A. Moore describes how to market innovations through the lifecycle in Crossing the Chasm.

Indeed began to donate refurbished MacBook Air and MacBook Pro laptops to graduates of the Family Self-Sufficiency program as well as to youth who achieve A/B Honor Roll and Perfect Attendance.

> It was really helpful to be on the same level as other classmates who had the advantage of internet at their house. I was finally able to keep up. It relieved a lot of the struggle and stress of applying to scholarships and colleges. And it allowed me to find my passion for science and technology—that wouldn't have happened if I didn't have internet.
> —Christeen Weir, who graduated as valedictorian of her class in 2016. She is the first in her family to attend a four-year university and now pursues a degree in biomedical engineering

The second phase of the Unlocking the Connection initiative focused on applying digital inclusion to fundamental barriers of resident self-sufficiency and quality of life. Initiatives related to energy efficiency, transportation, and citizen engagement relied on stipend-paid HACA residents to design and deliver programs that would help their neighbors leverage connectivity and earned devices. Digital Ambassadors developed specialties in smart city arenas:

- Energy Ambassadors learned to use digital tools to help residents use digital thermostats, reduce energy consumption, and monitor utility bills at one pilot site.
- Mobility Ambassadors helped residents learn to use digital tools to navigate transportation, share their transportation needs using online and face-to-face fora, and advocate for partnerships with digitally powered transportation companies.
- Smart City Ambassadors will develop expertise spanning digital inclusion, deliberative democracy, and advocacy for equitable solutions to smart city problems—from connectivity to transportation to AI.

Energy Ambassadors participated in a pilot program to install Nest thermostats at a property with free Google Fiber, and to teach residents how to use a smart phone to manage the thermostat while using Austin Energy's online tools to manage their utility bills. Programmable thermostats are often used in low-income housing to reduce utility bills, but they are largely unsuccessful because they are difficult to program. Note that digital thermostats often have a dial with a mechanical interface, while programmable thermostats rely on a visual and symbolic literacy familiar only to those who use computer keyboards. While digital thermostats can be user friendly, the installation technician must install the thermostat using a "Web 2.0"-like interface. The thermostat dial in effect becomes a mouse that "scrolls" through hierarchical menu options. Arrows indicate hierarchical menu choices—not "do you want me to turn?" as one staff person asked.

Lauded as "intuitive" by the tech industry and tech-savvy consumers, maintenance staff who often lack digital literacy skills (even if they own and use a smartphone) struggled to use the digital interface to install the thermostats. However, staff have necessary context. One maintenance staff person pointed out that the "self-learning" thermostat, which learns the occupants' schedules based on movement, would not work for the bedridden or residents with poor mobility, the same clients who said they had no need for an internet connection. Soon one of the residents shared that she appreciated her thermostat most because she no longer had to wait for her caregiver to navigate Austin traffic to adjust her thermostat. Her caregiver could adjust her thermostat remotely using a smartphone.

Transportation is a great barrier to social and economic mobility for low-income people. Austin Mayor Steve Adler declared 2016 the Year of Mobility. On the eve of a historic $720 million transportation bond, Next Century Cities awarded Austin Pathways a grant for a new mobility equity initiative, Smart Work Learn Play. Soon, ten HACA residents were stipend-paid Mobility Ambassadors to: 1) ensure that HACA residents learn to use digital tools to navigate transportation; 2) navigate for meaningful transportation partnerships, with rideshare and bikeshare companies; 3) advocate for the needs of public housing residents, through online and face-to-face citizen engagement. Mobility Ambassadors were trained by CapMetro staff to conduct transit adventures with other residents, using refurbished smartphones. Car2Go offered to provide training and reduced-cost minutes to HACA residents and Mobility Ambassadors evaluated the best use cases for carsharing, such as when a $3.00 Car2Go trip would enable a resident to work a whole shift instead of a partial shift.

Cities are increasingly using online tools in addition to public gatherings to gather citizen input. The Austin Transportation Department's Vision Zero initiative aims to eliminate traffic injuries and fatalities. Since low-income people are more likely to use active transportation and travel to and from work before or after daylight, the program is of special interest to Mobility Ambassadors. The Austin Transportation Department provided training to Mobility Ambassadors on how to use an online tool to record locations HACA residents felt were dangerous. Mobility Ambassadors then worked in HACA property computer labs and canvassed properties in areas with no data points, using tablets and smart phones to ensure HACA resident transportation safety concerns were represented. At one point, Mobility Ambassadors doubled the number of safety data points reported by residents in low-income zip codes.

The Transit Empowerment Fund invested in expanding Smart Work Learn Play to ensure that low-literacy Austin residents could use digital transportation planning tools to reduce the cost of transportation for the family. Mobility Ambassadors surveyed 240 HACA residents about the trips they take—and don't take—to determine the cost burden of transportation from different properties to find fresh food, get a flu shot or go to school. Thirty percent of respondents' reported trips related to healthcare so Mobility Ambassadors began to evaluate healthcare transportation options offered by digitally powered transportation providers.

To kick off National Digital Inclusion Week in 2017, HACA Mobility Ambassadors helped host a Mobility Equity Summit in Council Chambers at City Hall. Representatives of major transportation providers for the first time met with digital inclusion advocates and trainers to educate 40 HACA residents and staff about how to use digital tools to move about the city and advocate for their needs. For dozens of HACA residents, it was the first time they had been to City Hall.

More recently, Lyft has contributed to the Smart Work Learn Play fund to ensure that residents learn how to use all forms of digitally powered transportation, and that residents can use digital tools to successfully work as drivers, educators, and to maintain scooters and e-bikes. We believe micro-transit companies will depend on digital literacy in low-income neighborhoods both to sustain their workforces and to secure lower-margin revenue from lower-cost transportation products.

The 2019 cohort of Mobility Ambassadors will take on new challenges and expand their role as Smart City Ambassadors. Building on their knowledge of smart transportation, each Smart City Ambassador will be expected to develop subject matter expertise on an aspect of the City of Austin's smart city strategic plan. And each will be expected to participate in outreach, education or special projects related to an aspect of digital inclusion: affordable, robust broadband internet service; internet-enabled devices that meet the needs of the end user; access to digital literacy training; technical support and applications and online content designed to enable and encourage self-sufficiency, participation, and collaboration.

As residents adopt the internet and smart city technologies, they realize the need for multiple devices. A desktop at home is ideal for doing homework, watching the news or a free movie downloaded with a library card, and allows for low-cost storage of digital media. A laptop with offline interactive content enables a parent going to school to study or learn in the "in-between time," the time waiting for a child to complete an after-school activity or time used after a late shift to submit homework to a learning management system by midnight. A smart phone or tablet is required if a resident needs to use digitally powered transportation, coordinate a low-cost delivery of food or medicine, or simply to stay safe and in-touch with family on a bus ride or at work. The Austin Forum on Technology & Society, a non-profit that hosts monthly events to explore the impact of technology on society, began to collect used smart phones and tablets for HACA residents. These second generation and older devices, after being scrubbed by a third party, will enable hundreds of youth and seniors to access smart devices.

By 2018, 55% of HACA households had a free internet connection, up from 4% in 2014. Almost 1000 HACA residents had earned devices for their homes, having participated in more than 35,000 h of digital literacy training. A handful HACA resident Digital Ambassadors and Lab Apprentices had been hired by City of Austin agencies to help bridge the digital divide and educate Austin residents about new initiatives.

> It's for us. It's not just for them. It's for us
> —Jan Morgan, HACA Smart City Ambassador [22]

8   Next Steps and Questions to Ask

The overall understanding of an implementation program for the "Smart City" needs both technical skills of the hardware/software experts and the understanding of the social implications from social scientists. The "team" (and it more than one person) must be able to identify the needs that will be covered to any given project. Once a needs list is prioritized, the budget must be delineated.

The planning to address the needs must cover what will be the actual result and an idea of how to implement the plan. There will be an interplay between the social scientists and the hardware/software experts. What may seem trivial to the equipment experts may have serious challenges from the social perspective. Likewise, a simple request to address a social need might require extensive hardware and software implementation.

Smart City Maturity models should be reviewed to include dimensions of inclusion and equity, so that a wide range of stakeholders can engage more effectively in doing what they seek to do. Cities should adopt frameworks that prioritize inclusion and equity in smart cities projects and programs.

The following sections may help to define "smart" via several avenues of questioning involving municipal entities, public and private stakeholders, and community expectations.

8.1   Defining "Smart" via Digital Equity and Inclusion

- How smart can a city be if some are not connected? How can we define digital equity, given the lack of digital inclusion standards for broadband access, digital literacy, and device ownership and capacity?
- Does your community have a winning strategy to bridge the digital divide? Can you overcome the economics of broadband deployment, either in rural or low-income urban areas, so that broadband investment does not continue to flow into high-profit areas? What role can anchor institutions play? (see Schools, Health & Libraries Broadband Coalition)
- Can you document the state of broadband equity in your community? Could your coalition glean meaningful information from FCC Form 477 Census tract and block data, the U.S. Census American Community Survey, local survey data, library computer-user statistics, workforce development providers, social services staff, and affordable housing providers?
- What could your coalition do to build or improve broadband infrastructure in underserved and underconnected local communities? Could your city provide incentives, such as charging less for local rights-of-way access or providing new licensing options? Could private partners investing in smart city programs be required to make enforceable commitments?
- How could the responsibility to pay for technology adoption in various communities be shared by partners in diverse sectors such as education, workforce development, transportation, and healthcare?

8.2   Defining "Smart" via Partnerships and Coalition Building

- What role could coalition(s) play in your smart cities and communities project? What role could open source initiatives and movements, anchor institutions, corporate employee resource groups, non-profit organizations, and philanthropic organizations play? (see National Digital Inclusion Alliance, [2])
- Which banks, hospitals, and government agencies have strong business reasons to invest in equitable smart cities and communities projects? Banks' investment in community digital training or network access may now count as a qualifying activity for Community Reinvestment Act credit, which is something every U.S. bank needs to pass its federal regulatory reviews. (See: The Dallas Federal Reserve Bank's Closing the Digital Divide: A Framework for Meeting CRA Obligations.)

8.3   Defining "Smart" via Decision-making: Autonomy and Trust

- Do you have a personal and professional perspective on the role of diversity, equity, and inclusion in your smart cities and communities roadmap? Do you tend to make decisions from the perspective of the Guardian, the Advocate, or the Innovator? How will your coalition build and incorporate diverse perspectives?
- Given a smart cities and communities project, could you identify how these distinct variables, diverse perspectives; equality of opportunity and parity; and an inclusive, welcoming environment, affect your smart project goals?
- Do all the required organizations buy-in to the required planned implementation?
- Which banks, hospitals, and government agencies have strong business reasons to invest in equitable smart cities and communities projects? Banks' investment in community digital training or network access may now count as a qualifying activity for Community Reinvestment Act credit, which is something every U.S. bank needs to pass its federal regulatory reviews. See the Dallas Federal Reserve Bank's Closing the Digital Divide: A Framework for Meeting CRA Obligations.
- How could the responsibility to pay for technology adoption in various communities be shared by partners in diverse sectors such as education, workforce development, transportation, and healthcare?
- How much trust is required in the multi-dimensional systems you are building? What is the cost of built-in trust? What is the projected impact (cost, quality) of not building for trust? For example, how much trust is needed for marginalized residents to use telemedicine and telehealth tools for physical and mental health and wellness?
- Do all the required organizations buy-in to the required planned implementation?

References

1. Caird S, Hallett S (2018) Towards evaluation design for smart city development. J Urban Des 24(2):188–209. https://doi.org/10.1080/13574809.2018.1469402
2. National Digital Inclusion Alliance. https://www.digitalinclusion.org/
3. Lewis J, Severnin E (2017) Short- and long-run impacts of rural electrification: evidence from the historical rollout of the U.S. power grid. IZA Institute of Labor Economics, DP No. 11243. http://ftp.iza.org/dp11243.pdf
4. Fu H, Mou Y, Atkin D (2015) The impact of the telecommunications act of 1996 in the broadband age. Chapter 4 of Advances in Communications,Volume 8. A. Stavros, ed. ISBN: 978-1-61324-794-5. Nova Science Publishers, Inc.
5. Hovis J (2018) Broadband infrastructure [Video file]. Retrieved from https://www.c-span.org/video/?440451-1/hearing-focuses-broadband-infrastructure&start=7995&transcriptQuery=smart%20city
6. Anderson M, Perrin A (2018) Nearly one-in-five teens can't always finish their homework because of the digital divide. Pew Research Center, Washington, DC. https://pewrsr.ch/2JirZar
7. DePillis L (2017) Inequality among America's seniors some of the worst in developed world. CNN Money. https://money.cnn.com/2017/10/18/news/economy/elderly-income-inequality-oecd/index.html
8. Smyth C (2019) Millions of patients to see hospital doctors by Skype under NHS plan. The London Times. https://www.thetimes.co.uk/article/millions-of-patients-to-see-doctor-by-skype-under-nhs-plan-jj9fwmlk6
9. Staffing Industry Analysts (2018) $864 billion in revenue generated from US gig work. https://www2.staffingindustry.com/site/Editorial/Daily-News/864-billion-in-revenue-generated-from-US-gig-work-SIA-47628
10. Farrell D, Greig F, Hamoudi A (2018) The online platform economy in 2018: drivers, workers, sellers, and lessors. J.P. Morgan Chase & Co. Institute, New York. https://www.jpmorgan-chase.com/corporate/institute/report-ope-2018.htm
11. Talbot D, Hessekiel K (2018) Community-owned fiber networks: value leaders in America. Berkman Klein Institute for Society and the Internet, Cambridge. https://cyber.harvard.edu/publications/2018/01/communityfiber
12. Federal Communications Commission (2018) Fixed broadband deployment data from FCC form 477. https://www.fcc.gov/general/broadband-deployment-data-fcc-form-477
13. Marcus M (2017) AT&T accused of digital redlining in Detroit. Community Networks. https://muninetworks.org/content/att-accused-digital-redlining-detroit
14. National broadband plan. 4 Feb. 2019. Retrieved from https://en.wikipedia.org/wiki/National_Broadband_Plan_(United_States)
15. Campbeel-Dollaghan K (2018) Sorry, your data can still be identified even if it's anonymized. Fast Company. https://www.fastcompany.com/90278465/sorry-your-data-can-still-be-identified-even-its-anonymized
16. Przeybilovicz E et al (2018) A tale of two 'Smart Cities': investigating the echoes of new public management and governance discourses in smart city projects in Brazil. In: Proceedings of the 51st Hawaii international conference on system sciences. https://aisel.aisnet.org/hicss-51/eg/smart_cities_smart_government/2/
17. Kitchin R (2014) The real-time city? Big data and smart urbanism. GeoJournal 79:1. https://doi.org/10.1007/s10708-013-9516-8
18. Kahneman, D (2011). Thinking Fast and Slow. Farrar, Straus and Girroux. ISBN 9781429969352. https://us.macmillan.com/books/9781429969352
19. wzamen01 (2009). HP computers are racist [Video file]. Retrieved from https://www.youtube.com/watch?v=t4DT3tQqgRM
20. Chen BX (2009). HP investigates claims of 'Racist' computers. Wired Magazine. https://www.wired.com/2009/12/hp-notebooks-racist/

21. Rose A (2010). Are face-detection cameras racist? Time Magazine http://content.time.com/time/business/article/0,8599,1954643-2,00.html
22. National Institute of Standards and Technology (2018). NIST evaluation shows advance in face recognition capabilities. https://www.nist.gov/news-events/news/2018/11/nist-evaluation-shows-advance-face-recognition-softwares-capabilities
23. Miles A (2018) Digital inclusion program helps older and low-income austinites catch up with technology. KUT 90.5. http://www.kut.org/post/digital-inclusion-program-helps-older-and-low-income-austinites-catch-technology

Smart Responders for Smart Cities: A VR/AR Training Approach for Next Generation First Responders

George Koutitas, Scott Smith, Grayson Lawrence, and Keith Noble

Contents

1 Introduction to Training of First Responders

Mass casualty incidents are a national public health concern. Since 2012, there have been 14 natural disasters that have decimated cities with the financial costs estimated at $102.9 billion so far, not accounting for the three latest hurricanes [1]. Similarly, mass shootings have injured approximately 645 individuals, killed 203 people, and the USA accounts for 31% of public mass shootings worldwide [2, 3]. In one year, EMS responds to more than 14,000 events categorized as mass casualty incidents across the USA [4], and training for these events has become complex and a national initiative [5, 6]. Training and education—particularly using

G. Koutitas (✉)
Ingram School of Engineering, Texas State University, San Marcos, TX, USA
e-mail: george.koutitas@txstate.edu

S. Smith
Augmented Training Systems Inc., Austin, TX, USA

G. Lawrence
Department of Design, Texas State University, San Marcos, TX, USA

K. Noble
Austin – Travis County Emergency Medical Services, Austin, TX, USA

© Springer Nature Switzerland AG 2020
S. McClellan (ed.), *Smart Cities in Application*,
https://doi.org/10.1007/978-3-030-19396-6_3

technology—have been identified as key determinants of coordinated and successful response [7, 8]. In [9], the author suggests that the USA requires a more robust national response system and training should focus on enhancing response capabilities through improving "just in time" logistics and situational awareness.

Presently, emergency response training is not meeting the needs of growing prevalence of hurricanes and other large-scale incidents. For first responders, most training still follows the traditional exposition-type learning by providing theories and examples [10]. Evidence reported from hospitals [11], triage assessments [12], and EMT training in structural collapse scenarios and multi-agency response events [13, 14] suggests that emergency response training does not meet the needs of the growing prevalence of mass casualty incidents. Globally, emergency response to mass casualty situations has demonstrated that professionals do not feel equipped to address the needs of these high stress large-scale events [15]. A number of efforts have been studying effective ways to train first responders and medical personnel [16–20] and live simulations have been the most widely used strategy to prepare for these events [10, 19]. However, moulage or live training has several limitations including high cost, geographic barriers, limited feedback about performance, and it requires shutting down a number of stations to deliver the training [10].

The desire and budget allocation for first responder training has increased dramatically to over $150 million, with more participants signing up for classes on a national level, but budget allocations still restrict the type of classes and methods used [21, 22]. Continuous first responder training and education are critical components of professional development. Improving expertise is a complex process that involves several methods of instruction and instructional technology. Due to the nature of the training objectives, the training process requires frequent, often very expensive resources and high levels of inter- and intra-organization coordination. Additionally, it is extremely challenging to provide a multitude of scenarios that capture all training possibilities found within mass casualty or natural disaster events. Finally, and most important, some of the "live" training procedures expose first responders to risk of death and physical injury; indeed, 20% of fatalities are training related [22].

Another concern is that most training is focused on the basics, or "nuts-and-bolts," of rather than the more nuanced skills and expertise required to succeed on the incident ground. Even though training continues to be critiqued and updated, there remains a "need to develop training to achieve good habits in cue recognition plus the development of a greater appreciation of the nature of interactions between physical and cognitive processes on the incident ground" ([23], p. 203). Researchers also cite a lack of consistency in the delivery of training. For instance, Ron Cheves, former volunteer firefighter turned chief, argues "If you talk to 20 different fire departments, you will hear 20 different requirements for training and 20 different number of hours required to meet the minimum set by the individual department" ([24], para #1). Clearly, there is an urgent need to standardize and optimize the training programs offered to first responders.

2 AR/VR Technologies for Training

Augmented reality (AR) and virtual reality (VR) are considered as the technological foundation for the transformation of the training sector. Following the evolution of learning management systems (LMS) that leveraged cloud-based technologies to deliver a scalable and horizontal learning platform across disciplines, the AR/VR technologies are expected to provide a similar transformation but for the learning process that involve physical activities. Thus, the training aspect can be considered one of the most attractive use cases. An LMS is a software application that allows administrative users to administer, document, track, and access advanced reports of education courses and empowers the users with the ability to access educational courses anywhere and at any time. In a similar path, a training management system (TMS) can be considered a more customized LMS software that is used by training providers to help manage their training processes. The use of a TMS alleviates a lot of the back-office tasks and can be thought as a more corporate-level eLearning system.

LMS has been widely tested and evaluated and over the years it is found that such type of eLearning platforms improve student (or trainee) retention and engagement. The educational benefits of an LMS were showcased in [25]. In addition, mobile application of an LMS can improve engagement on the learning process [26]. It is clear that new cloud-based mobile technologies have changed the way people learn and train. For first responders training, LMS systems have been developed and used to provide content customized for the needs of EMS. In many cases, these LMS systems are based on state and/or city initiatives such as the Louisiana and Indiana first responders LMS systems. From the federal perspective, one of the most widely used distance learning training courses is Emergency Management Institute (EMI) https://training.fema.gov/emi.aspx and National Training and Education Division (NTED) https://www.firstrespondertraining.gov/frt/. EMI serves as the national focal point for the development and delivery of emergency management training to enhance the capabilities of state, local, and tribal government officials. NTED serves the nation's first responder community, offering more than 150 courses to help build critical skills that responders need to function effectively in mass consequence events. NTED primarily serves state, local, and tribal entities in 10 professional disciplines, but has expanded to serve private sector and citizens in recognition of their significant role in domestic preparedness.

One of the main drawbacks of existing first responder training is that it is still in "analog" phase. The training usually involves the physical presence of first responder in the training field and the trainer cannot access performance analytics of the trainee. Both the physical presence and analytics problems are partially addressed by LMS systems but there is one more drawback of existing solutions. This is the physical memory. The existing training material are based on 2D presentations, usually incorporating a lot of wording and some images making the gained cognitive and physical memory of the first responder limited. To address this problem, immersive training experiences that involve the human body have been developed using AR/VR tools.

Virtual reality (VR) is a technology that "isolates" the user from the real physical environment using a specialized headset with a large display. This display projects a virtual environment where the user can experience specific 3D scenarios. The virtual environment is based on 3D objects that are developed to satisfy the need of the experience. There are numerous VR headsets in the market each one presenting its own characteristics such as resolution, position tracking, eye tracking, hand tracking, and immersive depth.

On the other hand, augmented reality (AR) is a more advanced technology that empowers the user with the ability to interact with holograms which are overlaid on the physical environment. AR [27] is the perception of digital information—usually graphics, visually synchronized with objects and places in the physical world around the user. One of the most important differentiators of AR technology is the spatial mapping algorithm that is capable of mapping the physical space using depth cameras and sensors and position holograms on top or behind physical objects giving a "mixed reality" sense to the user. This concept is called occlusion. This technology empowers users with the ability to interact with the object and holograms, but also holograms to interact with the objects creating an enhanced experience. Thus, the user is able to navigate and interact with the real physical environment but also experience a virtual layer of holograms and 3D objects.

An interesting AR application for first responders is the integration of AR with internet of things (IoT). Triaging and patient health monitoring can be performed by integrating a sensor network of internet of things (IoT), deployed at a patient's body (i.e., wrist-bands or other wearables) [28] with an AR device to create a "4D" experience. The 4D experience provides real-time spatio-temporal visualization and allows the user (EMS personnel) to interact with the IoT network and patient in a highly intuitive fashion, improving patient care and improving EMS effectiveness during an event. AR has been used to quantify attention shifts in neurosurgical operations [29] and for enhanced surgical experiences [30, 31] as well as in nursing [32].

The use of AR/VR technologies for the training of first responders has gained a lot of momentum over the last years. One of the largest initiates is the enhanced dynamic geo-social environment (EDGE) tool was funded by the US Department of Homeland Security Science and Technology Directorate (DHS S&T) and the US Army Research Laboratory, and developed with input from first responders. The platform allows users to assume discipline-based avatars and role-play complex response scenarios. Researchers at the National Institute of Standards and Technology (NIST) aim to make virtual reality simulations more of a reality for first responders, enabling firefighters, law enforcement officers, and others to train for emergency operations and communications. In a parallel way, a combined AR/VR platform for the training of first responders has been used in the ambulance bus [33]. Federal, state level but also private corporate initiatives are now in full acceleration trying to provide a unified training platform that will allow first responders to experience immersive training and improve their performance.

The use of AR/VR technologies in the field of training for first responders is important since it can significantly reduce the costs but also improve training efficiency. Training expenses are usually related to the need of physical transportation of first responders to the training locations. If the AR/VR platform can provide an immersive training environment at the first responder premises, then the cost for training associated with the city level and the first responder itself can be significantly reduced. Another important benefit is the diversity of training environments. AR/VR is based on 3D objects and environments and this provides infinite degrees of freedom to a training designer to replicate existing disaster scenarios or even create new ones. In addition, the software-based training approach empowers the trainer and trainee with the ability to customize the training environment and also access analytics of the training processes, which are not available up to now. Based on the concept, "you cannot manage something that you cannot measure," it is expected that the data analytics of the training will provide significant insights to help improve the overall processes.

3   City of Austin AmBus: Introduction

To address large-scale events and or mass casualty response, many cities are deploying what is called an AmBus or ambulance bus. The AmBus for the city of Austin, Texas is shown in Fig. 1, and is one of only 13 in the state of Texas. Each AmBus can transport up to 20 stretchered patients at one time, while providing them much needed medical assistance along the way in the region. The vehicle may be used for

Fig. 1 AmBus at the city of Austin

a variety of missions, including rehabilitation, local or regional mass casualty incidents, and state deployments.

Currently, AmBus staff receive basic training consisting of a 20-min PowerPoint, and a quick walk through of the bus. All first responders have previous knowledge of what kind of equipment goes on an ambulance, and how to use that equipment. However, they are only given roughly an hour of training, that happens only once, in order to familiarize themselves with the AmBus [34]. The disaster medical response (DMR) team receives an extended 2-h training on top of this. Their training consists of the same PowerPoint, a quick walk through, and a scenario. The scenario consists of training activities like unloading and loading patients, and hooking patients up to vital sign monitors. Another scenario consists of a mock nursing home evacuation. The DMR receives a slightly more extensive training because they will be the ones deployed to long-term events, like Hurricane Harvey [34]. Once trained, it may be months before a member of the DMR team is deployed to a mass casualty event, and the only training they have to guide them may already be outdated.

4   A System Model for the Training of AmBus Using AR/VR

One of the main objectives of the developed training platform is to improve the training efficiency, the engagement of first responders in training, but also reduce the costs associated with their training. It is clear that there is an untapped market currently being addressed by outdated technologies. The proposed AR/VR technology follows a platform-based approach rather than a simple customized application. The training platform provides a reporting tool to the trainer so as to have a clear overview of the performance of the trainees. In addition, the platform offers the ability to customize and create various training tasks to capture a great diversity of real life disaster scenarios.

Since both virtual and augmented reality have their advantages and disadvantages, the platform offers the same type of AmBus training with both technologies. The first responder is able to choose the type of training he/she prefers or even perform both. The main advantage of the VR technology is that it has no constraints in terms of the design of the 3D environment. Theoretically, the user can be transferred in a virtual world and can perform any type of training that in some cases it might feel like a game instead of a training and this increases the engagement. For the purpose of our investigation, the Oculus Rift headset device was used that provides advanced touch controllers and spatial mapping capabilities. The VR headset was connected to a computer that was responsible to run the simulation, using high end graphics card and Windows 10 system. The VR simulation was designed in a specialized software called the 3D studio max and the development of the application in unity language.

On the other hand, the AR system provides an important benefit since it has the potential to stimulate the kinesthetic memory together with the cognitive memory of the user. The AR headset devices instead of having a display in front of the user's eyes have transparent lenses and all the holographic content is projected on the lenses. For our study we used Microsoft HoloLens AR device. One of the most important characteristics of AR is the mixed reality experience through occlusion and spatial mapping. The AR device can understand the physical environment and allow holograms to interact with it. The AR headset, similar to the VR headset, the holographic training environment of the AmBus was a 3D object input that with the use of spatial mapping technique was attached to the corridors of a building. This overlay provided the experience to the user that is actually inside the AmBus. The AR and VR user experience is presented in Figs. 2 and 3.

In addition, the AR engine enables an interaction with the basic elements of the AmBus such as drawers and medical equipment. In the developed application, the user is able to open/close drawers, visualize the various objects and key elements of the AmBus, and tap on specific objects to select or de-select them when necessary. This is an important feature that improves the user experience (UX) during the training process.

The AR/VR training platform incorporates three phases. These are the (a) explore phase, (b) training phase, and (c) reporting phase. The explore phase is the default scenario when the user launches the application. At this phase the AmBus environment and its important equipment are showcased together with labels to teach the user where each item is. An example is presented in Fig. 4.

In this untimed, zero-pressure mode, cadets can explore and study the environment as much as they need before training. At the training phase, which is the actual simulation of the physical test, cadets are given a series of tasks to complete on the virtual AmBus. For the purpose of our investigation, there was a set of 11 tasks to test cadet's memory recall of the AmBus systems. For example, the cadet is tasked

Fig. 2 Cadet studying the AmBus in the finalized AR

Fig. 3 Cadet studying the AmBus in the finalized VR

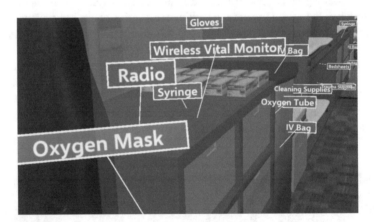

Fig. 4 The explore mode of the AR/VR system

with finding a specific object within the bus such as the medical kit, or the IV solution, or the radios. Each task is associated with a timer so the system can capture analytics of performance. To complete a task, the cadet should interact with the right object/equipment. A notification was used to present to the cadet about the correct or false selection of the object of the task. This notification is presented in Figs. 5 and 6.

The number of errors was also recorded for the insights of the reporting phase. The reporting phase is used by the trainer in order to obtain an overview of the performance of the first responders. The reporting phase presents the efficiency of the training for each cadet but also for the class. Together with the information of the efficiency engagement insights were reported such as the time spent on the training and the comparison of the performance compared to their cadets.

Fig. 5 Notification that indicates the correct completion of a task

Fig. 6 Notification that indicates the false selection of an object/equipment and thus unsuccessful completion of a task

5 Design Thinking for Training of FR

In order to better identify problems with the EMTs training, design thinking (DT) methodologies were employed in order to give the team focus, identify the various user needs and pain points for the project, and prioritize those pain points for the research. DT methodologies have been around for a number of decades, although the term was first coined by IDEO's Tim Brown and Roger Martin in the 1990s [35]. DT helps teams to become more user-centric in their problem solving, and aids in organizing and prioritizing large amounts of information, in order to make assumptions and suggest solutions to a user's problem. It also gives a framework to test those solutions with users, and iterate multiple times in order to design the most usable and implementable solution for the user. At its core, DT is user-centered, iterative, and relies on results of user-testing to evaluate design solutions.

Although there are different DT methodologies (i.e., IDEO, IBM Design, Google Sprint), they follow the basic six steps: empathize with the user, define pain points, ideate on multiple solutions, prototype chosen solutions, test those prototypes with the actual users affected by your designs, and implement those solutions deemed effective [35]. DT is an iterative process, meaning that the next version of the product goes through the same six-step cycle after implementation to make further improvements on the design.

As user experience (UX) and visual designers began to delve into ever-more complex problems, DT allows teams to synthesize large amounts of information and defines core problems to be solved, quicker and more effectively than before. IBM design, a major proponent of DT methodologies as they are applied to software products, restructured a large part of their business in 2013 to include DT in almost every facet of the software design process. A recent study commissioned by IBM found that the introduction of DT methodologies had increased team efficiency by 75%, reduced time to market by 2 times, and resulted in a return on investment of up to 300% over 5 years [36].

Our research team decided on a DT method called a design sprint [37] (not to be confused with an agile sprint, which is a software-development method). Design sprints originated out of Google's design ventures design process [37]. They consist of a number of DT activities, organized over 5 working days so the whole activity fits within a standard work-week, with the goal of testing a prototype with users at the end of the week. Sprints were chosen because they are designed to work well with interdisciplinary teams, with this research team being comprised of three different disciplines. As this project was an academic one—meaning the participants having other work obligations—the team conducted the sprint over a number of weeks. In addition, scheduling conflicts—including unexpected deployments—with Austin/Travis County EMS, meant that user-testing and prototyping process took a number of months to complete.

The research began the first sprint step of empathizing with the user, by collecting after action reports from AmBus personnel who have been deployed in the past, reviewing their current training, and developing a questionnaire to gather information from the potential users of our solution. From the information collected at this step, the team set out to identify pain points for AmBus personnel. Survey results noted that the majority (63.41%) of users would be advanced life saving (ALS) (43.90%) and basic life saving (BLS) (19.51%) EMTs. Respondents indicated that they had not had AmBus training for 1 year or more (81.82%) and 63.64% indicated a need for more training. In addition, write-in responses showed a pattern that AmBus personnel primarily were concerned with their performance with the AmBus layout, infrequency of training opportunities, loading of patients, and specific equipment training. While in interviews with EMTs, the utility of the paper triage tags used in the field was mentioned repeatedly.

To help define our objectives, the team reviewed the collected information and began to synthesize it into more digestible forms using DT methods of writing a Goal Statement, How Might We statements and card-sorting. An ideation session with the team resulted in targeting two possible solutions to the noted pain points. An AR tool to be used in the field to help triage patients and assist with patient loading

Fig. 7 AR Triage prototype

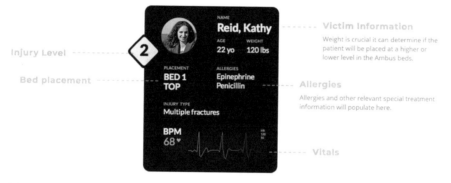

Fig. 8 Detail of digital triage tag and bed sorting location

as an enhancement to traditional paper-based triage tags, and a VR version of the AmBus in order to increase frequency of training on the equipment and enhance familiarity with the AmBus layout for both ALS and BLS personnel.

The initial AR prototype was designed to replace the inconvenient paper triage tags in the field. It consisted of using a Microsoft HoloLens to scan a code placed on a patient. While looking at the patient, the EMT would be provided with vital information (i.e., theoretically provided by a wearable biosensor placed on the body, as shown in Fig. 7).

Additional detail of a digital triage tag and bed sorting location are shown in Figs. 8 and 9. Figure 8 shows details provided by the biosensor tag about the patient, including conventional identification data (name, etc.) as well as an indication of injury level, bed placement, allergic reaction data, and vital signs such as blood pressure and heart rate. Figure 9 provides an overview of the location of the patient inside the AmBus, including suggestions for placement based on critical information such as injury level and weight.

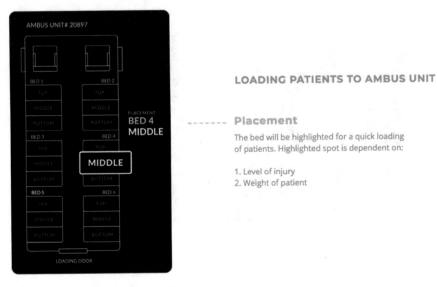

Fig. 9 Detail of digital triage tag and bed sorting location

The first VR prototype was created by taking 360° photos of the AmBus. The photos were taken of the entire length of the bus interior, and of each drawer in both the open and closed positions. These photos were then imported into an online tool called InstaVr (https://www.instavr.co/) which allowed for hotspots to be created throughout the bus (see Fig. 10). These hotspots were linked together in order to allow the user to virtually "walk" through the bus and open each of the drawers to view inside them. This interactive prototype was deployed on a smartphone using an inexpensive phone VR headset. To explore the AmBus, the user moved a fixed cursor dot in the middle of their view to one of the hotspots, leaving the cursor on the spot for a short time, which then would switch out the original photo with the new photo.

Testing both the AR triage and VR AmBus prototypes with a small number of EMTs and their instructors indicated a preference for the VR AmBus prototype. Secondary prototypes were created in both AR and VR of the AmBus interior. The visual design team created 3D rendered models of the AmBus and individual medical items to scale. In some cases, whenever stock 3D models of medical items were available, they were purchased to save time. These models were shared by both the AR and VR version of the AmBus. These prototypes were rudimentary, only allowing for the user to walk through the bus and open one or two drawers filled with items (see Figs. 11 and 12).

User-testing the second round of prototypes with the EMTs and their instructors validated the concept further and suggested features and improvements from EMT test subjects were collected. These included wanting interface improvements, a need to collect data on how individual test subjects.

Many of the suggestions from the EMTs incorporated into the next AR and VR versions of the prototype designed for validation testing. Two modes were added to

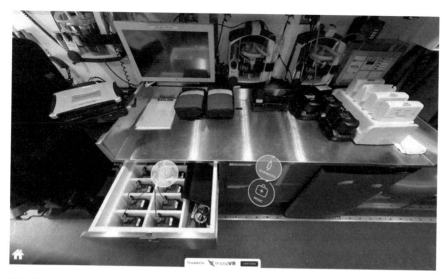

Fig. 10 AmBus 360° interactive prototype

Fig. 11 User-testing the AR version with EMT professionals

the prototype: a challenge mode and a learning mode. Learning mode allowed the user to casually explore the AmBus with no solid objective. Challenge mode, on the other hand, tasked the user to find certain objects or groups of objects, timed their performance, and recording number of errors.

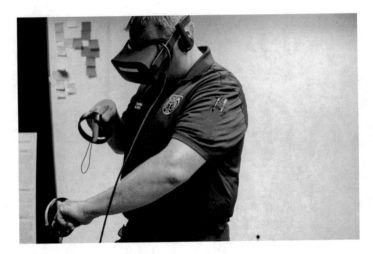

Fig. 12 User-testing the VR version with EMT professionals

6   Findings on Pilot Training

Validation testing of both the AR and VR versions of the AmBus was completed over 2 weeks in November 2018. Testing was administered at the police and fire training facility as well as the fire department headquarters in Austin, Texas.

6.1   *Methodology*

New recruits were used as test subjects, as they were not previously familiar with the AmBus and its layout. The recruits were split into three groups of 10 each: traditional, AR, and VR. The traditional group only received the current training consisting of a PowerPoint presentation and 1-h orientation of the real AmBus. Both the AR and VR groups were given the PowerPoint presentation, but not allowed the 1-h orientation in the real AmBus. Instead, AR and VR groups were allowed to use their assigned simulations as many times as they wished for 1 week. They were asked to use the virtual simulation a minimum of 3 times in the week and asked to use it in the morning of the final test. All three groups were asked not to discuss their experiences with each other during the testing procedure.

At the end of the week, all three groups were tested on the real AmBus. Their test consisted of finding various items on the bus, where they were timed by an EMT commander and assistant proctoring the test. The number of errors was also recorded by the proctors. After their test, participants were asked to fill out an online questionnaire designed to ascertain how they felt about the training and to give them opportunity to point out errors or give suggestions on further improvements.

6.2 Results

Of the 30 cadets signed up to participate in the training, 3 did not participate in the AR and 2 did not participate in the VR. Their scores in the final test were removed from the findings.

The VR training resulted in a 46% reduction in time-on-task and a 29% reduction in errors, while the AR training resulted in a 10% reduction of time-on-task and a 34% reduction in errors.

A total of 15 participants responded to the follow-up questionnaire. 90.91% found both AR and VR simulations were somewhat or extremely easy to navigate. 90.91% reported that identifying objects in the bus were somewhat or extremely easy. 81.82% felt that after AR or VR training, they were somewhat or extremely confident they could find items in the real bus if asked. 72.73% felt that no matter how many times they used the training, that it was helpful. 90.9% said they would be likely to choose a VR or AR training of another subject in the future. 81.82% felt that they were prepared for the test in the real AmBus after taking the AR and VR training. When asked if they preferred the VR/AR or traditional training, 36.36% had no preference, 27.27 either preferred or strongly preferred traditional training, and 36.36% either preferred or strongly preferred AR/VR training. Some participants reported discrepancies between the AR/VR and the real AmBus.

7 Conclusions

Virtual reality and augmented reality strategies (VR/AR) can address the above training gaps and integrate the physical and cognitive processes required in the first responder training. Some believe this approach is a more cost-effective way to address the training needs of first responders. Over the last decade, VR/AR has repeatedly been shown to be an effective training tool in diverse fields like surgery, combat, and treatment of various psychological conditions [38, 39]. Moreover, the technology has developed to the point where it is affordable to incorporate in training efforts. Virtual and augmented reality technology is increasingly available in mainstream applications, with barriers to cost and accessibility overcome by requirements for a commercial market. Currently, various affordable (ranging from $99 to $766) virtual reality platforms have been released to the general public and for $99 the Samsung Gear VR ($99) is adaptable to any Samsung smartphone. The low cost and high availability of VR technologies will increase the utility of the methods discussed here, including standardization in the training of first responders.

First responders are typically trained in the classroom and the field. Unfortunately, it is impossible with the traditional approach to expose first responders to the breadth and depth of scenarios they may face in a live emergency event. For budgetary, logistic, and safety reasons, departments must limit field training to the most common and important scenarios while keeping the risk of injury low and the costs within budget. VR/AR has the ability to bridge the gap between classroom instruction and live training. It safely immerses first responders into real-world experiences, thus expanding the breadth and depth of scenarios to which they are exposed. VR/AR also has the distinct advantage over "real world" training by isolating and enabling a focus on first responder judgment, decision making, and other responses when under stress. Deficits in cognitive recall, situational awareness, and decision making in high risk situations can be addressed.

The solutions mentioned in this chapter can augment and improve first responder safety, performance, and confidence by adding scenarios potentially difficult to replicate in real-world training; bringing "stress" closer to the trainee without putting them physically at risk. It will expand opportunities for departments to access cost-effective, safe ways to evaluate the decision making and psychological readiness of their teams. It can also be used to identify and train new recruits. And, unlike expensive systems requiring large up-front investments, the proposed systems will be affordable, portable technologies with tested curriculum and policies that can be adopted by departments large and small. While being deployed the AmBus system will develop additional user interfaces through haptic feedback that will be informed from health related IoT devices, improve patient care process with new technologies, and reduce the required time spent per patient.

Acknowledgements The team would like to thank Commander Noble from Austin/Travis county EMS and our contacts at the City of Austin: Marbenn Cayetano and Dr. Ted Lehr; for without their help, this project would not be possible. We would like to also acknowledge the important feedback and research of the faculty team and researchers Dr. Vangelis Metsis and Dr. Mark Trahan. Finally we would like to thank the student team: Clayton Stamper, Jose Banuelos, James Bellian, Dante Cash, Elija Gaytan, Victoria Humphrey, Shivesh Jadon, Chloe Kjosa, Lorena Martinez, Samantha Roberts, Kayla Roebuck, Chaitanya Vyas, and Shashwat Vyas for their hard work and dedication.

References

1. National Centers for Environmental Information (2017) Billion-dollar weather and climate disasters. https://www.ncdc.noaa.gov/billions/events/US/2012-2017
2. CNN (2017) Deadliest mass shootings in modern US history fast facts. CNN News. http://www.cnn.com/2013/09/16/us/20-deadliest-mass-shootings-in-u-s-history-fast-facts/index.html
3. Lankford A (2016) Public mass shooters and firearms: a cross-national study of 171 countries. Violence Vict 31(2):187–199
4. Schenk E et al (2014) Epidemiology of mass casualty incidents in the United States. Prehosp Emerg Care 18(3):408. http://www.tandfonline.com/doi/full/10.3109/10903127.2014.882999
5. Guha-Sapir D, Hoyoise P, Below R (2016) Annual disaster statistical review 2015: the numbers and trends. Centre for Research on the Epidemiology of Disasters. https://reliefweb.int/sites/reliefweb.int/files/resources/ADSR_2015.pdf
6. Schmidt MS (2014) F.B.I. confirms a sharp rise in mass shootings since 2000. New York Times. https://www.nytimes.com/2014/09/25/us/25shooters.html

7. Yin H, He H, Arbon P, Zhu J (2011) A survey of the practice of nurses' skills in Wenchuan earthquake disaster sites: implications for disaster training. J Adv Nurs 67:2231–2238. https://doi.org/10.1111/j.1365-2648.2011.05699.x

8. Heinrichs WL, Youngblood P, Harter P, Kusumoto L, Dev P (2010) Training healthcare personnel for mass-casualty incidents in a virtual emergency department: VED II. Prehosp Disaster Med 25(5):424

9. Carafano J (2017) Preparing responders to respond: the challenges to emergency preparedness in the 21st century. http://www.heritage.org/homeland-security/report/preparing-responders-respond-the-challenges-emergency-preparedness-the#pgfId-1070145

10. Stansfield S, Shawver D, Sobel A, Prasad M, Tapia L (2000) Design and implementation of a virtual reality system and its application to training medical first responders. Presence Teleop Virt 9(6):524–556

11. Higgins W, Wainright C III, Lu N, Carrico R (2004) Assessing hospital preparedness using an instrument based on the Mass Casualty Disaster Plan Checklist: results of a statewide survey. Am J Infect Control 32(6):327–332

12. Lerner E, Cone D, Weinstein E, Schwartz R, Coule P, Cronin M, Hunt R (2011) Mass casualty triage: an evaluation of the science and refinement of a national guideline. Disaster Med Public Health Prep 5(2):129–137. https://doi.org/10.1001/dmp.2011.39

13. Fernandez AR, Studneck JR, Margolis GS, Crawford JM, Bentley MA, Marcozzi D (2001) Disaster preparedness of nationally certified emergency medical services professionals. Acad Emerg Med 18(4):403–412

14. Archer F, Seynaeve G (2007) International guidelines and standards for education and training to reduce the consequences of events that may threaten the health status of a community. A report of an Open International WADEM Meeting, Brussels, Belgium, 29-31 October, 2004. Prehosp Disaster Med 22(2):120–130. https://www.ncbi.nlm.nih.gov/pubmed/1759118

15. Sapp RF, Brice JH, Myers B, Hinchey P (2010) Triage performance of first-year medical students using a multiple-casualty scenario. Paper exercise. https://www.researchgate.net/publication/44807700_Triage_performance_of_first-year_medical_students_using_a_multiple-casualty_scenario_paper_exercise

16. Parrish AR, Oliver S, Jenkins D, Ruscio B, Green J, Colenda C (2005) A short medical school course on responding to bioterrorism and other disasters. Acad Med 80(9):820–823

17. Kaji A, Coates W, Fung C (2010) A disaster medicine curriculum for medical students. Teach Learn Med 22(2):116–122. https://doi.org/10.1080/10401331003656561

18. Deluhery M, Lerner B, Pirrallo R, Schwartz R (2011) Paramedic accuracy using SALT triage after a brief initial training. Prehosp Emerg Care 15(4):526–532. https://doi.org/10.3109/10903127.2011.5698522011

19. Ingrassia PL, Prato F, Geddo A, Colombo D, Tengattini M, Calligaro S, La Mura F, Franc JM, Corte FD (2010) Evaluation of medical management during a mass casualty incident exercise: an objective assessment tool to enhance direct observation. J Emerg Med 39(5):629–636

20. National Domestic Preparedness Consortium (2017) NDPC 2016 annual report. FEMA. https://www.ndpc.us/pdf/NDPC.Annual.Report.2016.pdf

21. Pugh B (2015) Does the United States' first responder training program improve national preparedness? https://www.interagencyboard.org/system/files/resources/Training%20and%20Preparedness.pdf

22. U.S. Fire Administration (2016) Fire fighter fatalities in the United States in 2016. https://www.usfa.fema.gov/downloads/pdf/publications/ff_fat16.pdf

23. Ash J, Smallman C (2010) A case study of decision making in emergencies. Risk Manage 12(3):185–207

24. Cheves R (2012) Volunteer firefighters: can there be too much training? https://www.carolinafirejournal.com/Articles/Article-Detail/ArticleId/2483/Volunteer-Firefighters-Can-there-be-too-much-training

25. Ouadoud M et al (2017) Educational modeling of a learning management system. In: 2017 international conference on electrical and information technologies (ICEIT)

26. Hung P et al (2015) A study on using learning management system with mobile app. In: 2015 international symposium on educational technology (ISET)
27. Yuan Y (2018) Paving the road for virtual and augmented reality. IEEE Consum Electron Mag 7(1):117
28. Khan SF (2017) Health care monitoring system in Internet of Things (IoT) by using RFID. In: International conference on industrial technology and management (ICITM)
29. Léger E, Drouin S, Collins LD, Popa T, Kersten-Oertel M (2017) Quantifying attention shifts in augmented reality image-guided neurosurgery. Healthc Technol Lett 4(5):188–192
30. Sutherland C (2013) An augmented reality haptic training simulator for spinal needle procedures. IEEE Trans Biomed Eng 60(11):3009
31. Daher S (2017) Optical see-through vs. spatial augmented reality simulators for medical applications. In: 2017 IEEE virtual reality (VR)
32. Wullera H (2017) Augmented reality in nursing: designing a framework for a technology assessment. International Medical Informatics Association
33. Koutitas G (2019) A virtual and augmented reality platform for the training of first responders of the ambulance bus. In: 12th international conference on pervasive technologies related to assistive environments, 5–7 June 2019, Rhodes, Greece
34. Nobel K (2018) Personal interview
35. Gibbons S (2016) Design thinking 101. https://www.nngroup.com/articles/design-thinking/
36. Forester Research. https://www.ibm.com/design/thinking/static/media/Enterprise-Design-Thinking-Report.8ab1e9e1.pdf
37. Knapp J, Zeratsky J, Kowitz B (2016) Sprint: how to solve big problems and test new ideas in just five days. Simon & Schuster, New York
38. Sniezek J, Wilkins D, Wadlington P (2001) Advanced training for crisis decision making: simulation, critiquing, and immersive interfaces. In: Proceedings of the 34th annual Hawaii international conference on system sciences, 6 Jan 2001, Maui, HI, USA
39. Patterson R, Pierce B, Bell HH, Andrews D, Winterbottom M (2009) Training robust decision making in immersive environments. J Cogn Eng 3(4):331–361

Part II
Public Safety and Policy Issues

Smart Transport

Michael Brown

Contents

1 Introduction

Smart transport is a vital system within a smart city. Efficient and safe movement of people and goods throughout a city is at the heart of what makes a city "smart." However, enabling smart transport within a city is much easier said than done. Technologies are being introduced at a much faster rate than cities can adopt and city officials around the world are grappling with how to fund the deployment of smart transportation technologies while their existing infrastructure is underfunded and crumbling.

Even in the face of these challenges, we mustn't give up as there is too much at stake. The World Health Organization reports over 1.25 million traffic related

M. Brown (✉)
Southwest Research Institute, San Antonio, TX, USA
e-mail: michael.brown@swri.org

© Springer Nature Switzerland AG 2020
S. McClellan (ed.), *Smart Cities in Application*,
https://doi.org/10.1007/978-3-030-19396-6_4

deaths worldwide annually [1]. Imagine the calamity and public outcry if a Boeing 737 carrying 143 passengers crashed once per hour somewhere around the world. Yet this is exactly the death rate we are facing on our roads every day. On top of the deaths and injuries caused by traffic accidents, we are struggling with mobility as residents of cities worldwide are stuck in congestion for multiple days of time annually [2]. This congestion negatively impacts cities both economically and environmentally.

Smart cities work to enable intelligent transportation systems (ITS) by piloting and deploying innovative solutions to these challenges. ITS is undergoing somewhat of a revolution due to emerging technologies such as connected vehicles, automated vehicles, and enhanced decision support systems. Information and communication technologies (ICT) are rapidly changing what is possible and opening up new opportunities for cities to leverage these solutions to enable smart transport.

2 Applications for Smart Transport

There are a wide variety of applications that a smart city can utilize to enable smart transport. These range from safety applications that allow vehicles to communicate with other vehicles and infrastructure to environmental and mobility applications such as transit signal priority. These applications also range in complexity and maturity. Some applications have only been identified and defined. Others have been deployed, tested, and refined. The following sections provide an overview of some example smart transport applications that a smart city can leverage.

2.1 Emergency Electronic Brake Lights (EEBL)

The EEBL application is a prime example of why vehicle-to-everything (V2X) communication is critical to realizing safety benefits. When a vehicle experiences a hard-braking event (>0.4 G deceleration), it will transmit a basic safety message (BSM) to nearby vehicles and infrastructure with an event flag set that indicates the presence of the hard-braking event. This allows surrounding vehicles to detect a hard-braking scenario from a vehicle that cannot be "seen" directly using onboard sensors. An EEBL scenario is depicted in Fig. 1, where the blue car ahead of the bus is quickly decelerating, and transmitting to nearby vehicles a BSM with an event flag indicating a "hard-braking" condition. The white car cannot directly "see" the blue car's context or braking event due to occlusion by the intervening bus.

While this may seem purely like a vehicle-to-vehicle (V2V) communication scenario on the surface, it is actually quite important to a smart city. With properly placed infrastructure to communicate with these connected vehicles, municipal

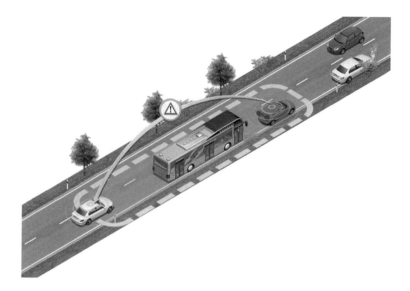

Fig. 1 Example of a basic safety message in a hard-braking event scenario

datacenters will also be able to receive the BSMs and monitor the data for detecting and responding to incidents as well as determining areas in which "hard-braking" events are common. In this fashion, cities can leverage such data to perform root cause analysis and possibly improve both the digital and physical infrastructure assets.

2.2 Queue Warning (Q-WARN)

Once a smart city has deployed roadside units (RSUs) in strategic locations, it can then use its communication backbone to notify upstream traffic of downstream hazards, or traffic queues. An example of a traffic queue is shown in Fig. 2 where slow or stopped vehicles (label 1) have queued up ahead due to some external problem. In the figure, RSUs aggregate slowed vehicle positions and provide context information to approaching vehicles (label 2).

This queue can be detected via several mechanisms including the BSMs that are transmitted from the vehicles and infrastructure-based sensing such as radar or cameras. A combination of these mechanisms provides a more robust queue detection mechanism which can reduce false positives. The warning of the upcoming queue can also be accomplished using a variety of communication mechanisms such as dedicated short-range communications (DSRC) [3], cellular vehicle-to-everything (C-V2X) [4], and/or 4th generation wireless telephony (4G LTE) [5].

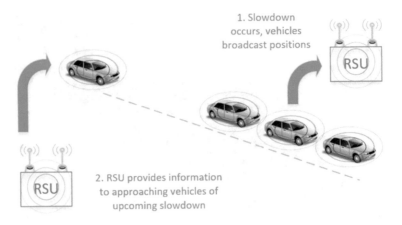

Fig. 2 An example of a traffic queue and Q-WARN system

2.3 Reduced Speed Zone Warning (RSWZ)

Construction is a reality for even the smartest of smart cities. Cities are constantly trying to maintain and repair their roadways while simultaneously trying to keep traffic moving. Work zones are difficult for AVs to navigate as the vehicles often depend on high definition maps to localize and navigate, and these maps are not updated at a frequency that would accurately reflect the work zone which can change frequently.

This is a great example of how a smart city can help to enable AVs that are not able to sufficiently deal with work zones. As part of a city's digital infrastructure, up-to-date maps can be transmitted to vehicles approaching a work zone to let it know about speed limit changes, lanes that are closed, and the appropriate route to take to navigate the work zone. An example of a construction zone with RSWZ is shown in Fig. 3. In the figure, the RSU broadcasts "reduced speed" warning signals to approaching cars.

Maintaining up-to-date maps of work zones is not an easy task. As a result, communications infrastructure will play an increasingly important role in monitoring traffic flow and communicating this information to upstream vehicles. This can help with work zones that are changing frequently or are very short-lived, and can help with situations such as temporary obstacles in the roadway.

2.4 Cooperative Situational Awareness

Breakthroughs over the last decade in the areas of deep learning and increased computational capabilities have led to a transition of intelligence in ITS to the edge.

The deployment of intelligent situational awareness hubs at key intersections allows vehicles to see pedestrians that are not detectable with sensors and to see

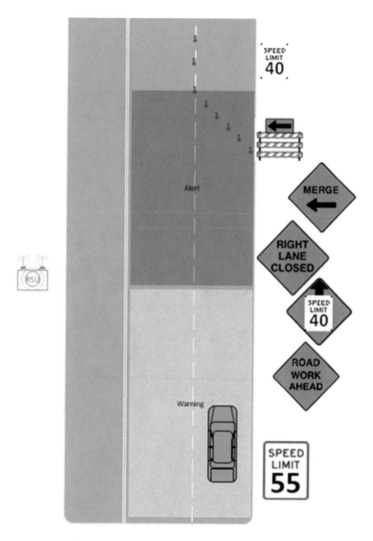

Fig. 3 Depiction of a reduced speed zone warning

around corners. This also allows complex computation to take place immediately where the data is received at the roadside and then information extracted from that data is exchanged with a traffic management center (TMC). This is in stark contrast to the traditional means of bringing all data back to the TMC and then dealing with it there. This capability can enable cooperative situational awareness for all users at the edge whether it is pedestrians, bicyclists, or vehicles. Figure 4 provides a descriptive perspective of cooperative situational awareness in a metropolitan area. In the figure, a situational awareness system located in/near a busy intersection aggregates information from cars, pedestrians, buses, and other transportation systems and communicates useful situational context to user systems.

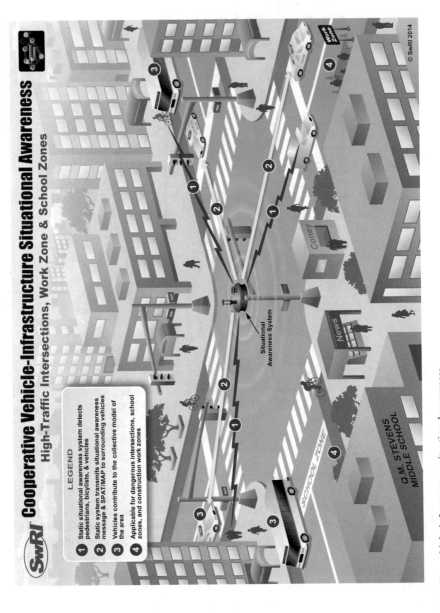

Fig. 4 Cooperative vehicle-infrastructure situational awareness

2.5   Integrated Corridor Management Systems

As ITS deployments become more mature and widespread, there is a growing need to augment these systems with decision support systems that regions can use to optimize the transportation network across modes. This capability is called integrated corridor management systems (ICMS) and is the wave of the future for smart transportation.

An ICMS collects data from a variety of public and private sector sources along a corridor, uses advanced modeling to predict the future state of the corridor if various actions are taken, and then takes the best course of action to optimize the flow of travelers along that corridor. Figure 5 provides an example of an ICMS in action. In the figure, accident detection and lane closures are used to redirect incoming traffic to alternate routes via connected-vehicle broadcast messages.

For example, consider the case shown in the following image in which a major accident has occurred on the freeway and traffic begins to back up significantly along the corridor. In an ICMS, the decision support system (DSS) will monitor the flow of traffic along the freeway and arterials, predict the future state of various potential scenarios (including rerouting traffic along arterials), then implement the best course of action. In this case the DSS routes traffic along the arterials while interfacing with the arterial signal systems to optimize the signal timing to allow the most traffic to flow along those arterials.

Following the event, the DSS can compare the impacts of the implemented response plan with the predicted impact. Machine learning techniques along with traffic engineering expertise can be utilized to tweak the capabilities of the system such that there is continual improvement in the ICMS in response to future events. This capability can be extended across many other modes including rail and buses.

2.6   Powertrain Optimization

The applications discussed so far illustrate the ability of smart transport technology to improve safety and mobility of travelers throughout the transportation network, but what about environmental impacts and fuel efficiency? There is a growing amount of research into this very question and it leverages the early work of the applications for the environment: real-time information synthesis (AERIS) program [6] along with other USDOT initiatives. ARPA-e (the DOE's advanced research projects agency) has initiated a $30M program investigating various techniques that would combine to enable CAVs to achieve at least a 20% fuel efficiency gain by optimizing the powertrain based on connected vehicle data. Imagine if your vehicle could get a preview of the traffic flow ahead and then optimize its powertrain accordingly. For example, the vehicle could more intelligently traverse a corridor based on the signal timing and other surrounding vehicles. Another example is the ability to optimize the power split in hybrid engines to utilize more battery and less fuel.

Fig. 5 Integrated corridor management system showing traffic diversions into alternate routes in response to an accident on the freeway

This capability is generating a lot of excitement not only in the passenger vehicle market but also in the heavy-duty vehicles. Large trucks and buses consume a significant amount of fuel and even a small savings can reduce their CO_2 emissions while simultaneously saving them significant amounts of money. Fuel is often the largest O&M cost for a fleet operator.

3 Enabling Technologies

So far, we've explored why smart transport is so vital to a smart city and what applications can be leveraged to improve transportation safety, mobility, and environmental impacts. Now let's look at "How?" What makes these applications tick and how can a city leverage these technologies without making investments in something that is quickly obsolete?

3.1 Navigating the Buzzwords

There are so many acronyms and buzzwords related to smart transport that it makes your head spin, e.g., ICT, IoT, 4G-LTE, DSRC, 5G, C-V2X, blockchain, cybersecurity, Wi-Fi, Li-Fi, OCC, NGV, 6G, deep learning, etc. The list is never-ending.

3.2 The Internet of Things

Let's begin to break these technologies down by starting at the highest level with communications systems architecture. For decades, smart transport systems have employed a strategy in which each agency deploys a certain amount of field equipment along the roadways and builds the network infrastructure to communicate with each device. Data from these field devices is collected and rolled up at the "center" level which typically refers to the building(s) that host the servers and communications equipment connecting the computer networks to the field networks.

This is how a traffic management center communicates with the cameras, message signs, traffic sensors, environmental sensors, and traffic signals that it maintains. The protocols used for this center-to-field (C2F) communication is sometimes proprietary but most often standardized in a National Transportation Communications for ITS Protocol (NTCIP) standard [7]. These protocols utilize a variety of existing communications media including fiber, cellular, Bluetooth, Wi-Fi, etc. The NTCIP C2F protocols are often based on older standards, such as the simple network management protocol (SNMP) [8]. Data that is collected from each center can be exchanged with other centers via the NTCIP center-to-center (C2C) protocol which

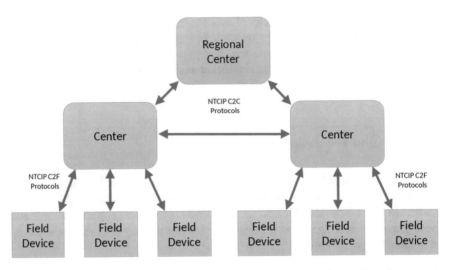

Fig. 6 Architecture showing how traffic management centers communicate with each other and with field devices

utilizes simple object access protocol (SOAP) [9] exchanging messages formatted in extensible markup language (XML) [10] compliant with the traffic management data dictionary (TMDD) [11]. This architecture is shown in Fig. 6, where the TMC locations communicate using "center to center" protocols, and field devices communicate with the TMCs using "center to field" protocols.

The term internet of things (IoT) simply refers to everything being connected to the Internet. Internet protocol version 6 (IPv6) allows an enormous set of available addresses such that even the tiniest of devices can have its own IP address and be accessible by any other device or system around the world. This increased addressing capability along with more and more communications bandwidth is facilitating a shift in communications systems architecture from this hub-and-spoke type model to a much more decentralized model. This architecture will realize itself soon by allowing systems to "share" resources such as ITS field equipment. For example, imagine a scenario in which a traffic management center has deployed traffic sensors along a corridor up to the point where another center has deployed its traffic sensors. To be able to calculate the travel time from one city to the other, a city must use the data collected from its traffic sensors along with data from the other center's sensors transmitted via the NTCIP C2C protocol. In a more decentralized architecture, each center can access the other centers' sensors directly to calculate the travel times that it needs to convey to those traveling along the corridor. A decentralized, interconnected communications architecture is shown in Fig. 7, where two field devices near the boundary of TMC centers communicate with both TMC sites.

This is just the tip of the iceberg, as sensors can be accessible to each center as needed allowing everything to communicate with everything. Furthermore, the

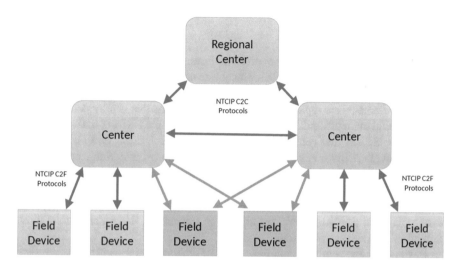

Fig. 7 Decentralized architecture where field devices are connected to multiple TMC sites

sensors are extending beyond just traditional traffic and environmental sensors and now even include mobile elements in the transportation system such as vehicles, bicycles, and pedestrians.

3.3   Vehicle-to-Everything

Modern vehicles are rolling sources of rich sensor data and can communicate this data with infrastructure as well as directly with other vehicles and road users. This is called vehicle-to-everything (V2X) communication and includes vehicle-to-infrastructure (V2I), vehicle-to-vehicle (V2V), and vehicle-to-pedestrian (V2P) communications. This capability extends the traditional intelligent transportation system from one that communicates with ITS field equipment to one that can collect data directly from vehicles and can communicate information directly to vehicles.

The communications media used for V2I communications can range from satellite and cellular for information that is not time critical to dedicated short-range communications (DSRC) and cellular V2X (C-V2X) for more time critical safety information that is localized. V2V communications is typically low-latency, small packets for higher reliability using DSRC or C-V2X. There are emerging technologies that utilize optical camera communications (OCC) which is a form of visible light communication (VLC) that modulates LEDs at a frequency higher than that visible to the human eye to transmit data from one source to another. Also, the evolution of DSRC is resulting in the development of next generation vehicular (NGV) communications which will support much higher data rates and packet reliability than DSRC.

3.4 Security

For a smart transport system to be effective, it must be secure. Elements within the transportation system must be able to trust information from other sources and must be resilient to hackers. To implement this security, messages that are exchanged with other elements are typically signed with a certificate that is provided by a security credential management system (SCMS). Currently the USDOT makes available a prototype SCMS that can be used for federal research pilots and there are also commercial alternatives emerging to support a full deployment. Some communications may contain personally identifiable information (PII) and often these data flows are also encrypted to ensure privacy. Other mechanisms are in place to ensure privacy by rolling over identifiers throughout the communications stack to ensure that vehicles or pedestrians are not able to be tracked for times longer than those critical to the safety applications.

Besides just the communications security of the transportation network, physical security of the ITS network is essential. Research that is combining all aspects of transportation system security is being conducted to develop guidance for state and local transportation agencies on mitigating the risks from cyber-attacks on the field side of traffic management systems (including traffic signal systems, intelligent transportation systems, vehicle-to-infrastructure systems (V2I), and closed-circuit television systems) and, secondarily, on informing the agency's response to an attack [12].

3.5 Standards

Standards are the key to enabling interoperability and system longevity in smart transport. We touched a little on NTCIP standards in communications between TMCs and with the field equipment but there are many other standards that have been developed or are being developed by standards development organizations (SDOs) for smart transport. Some of the major SDOs involved include:

- Institute of Electrical and Electronics Engineers (IEEE) [13]
- Society of Automotive Engineers (SAE) [14]
- International Organization for Standardization (ISO) [15]

Despite the significant amount of standards work that has been done to date in this area, there remains a lot of work to be done in the future. These standards will continue to evolve and adapt to allow new technologies to be introduced into our transportation system. The relationships between the elements of a smart transport system and the standards that facilitate interoperable communications between them are illustrated by the diagram in Fig. 8, which was created by the V2I deployment coalition [16]. The figure depicts a "logical view" of the standards context prescribed by the American Association of State Highway and Transportation Officials (AASHTO) [17].

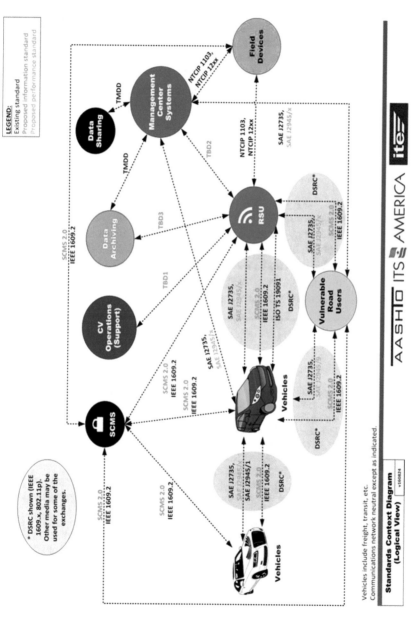

Fig. 8 Logical view of the AASHTO standards context

In the USA, several of the communications standards depicted depend on a 75 MHz band of wireless spectrum (5850–5925 MHz) that was set aside in 1999 by the FCC for ITS safety. This spectrum is currently under review by the FCC as the cable companies are pushing to have it opened for increased Wi-Fi access. In parallel, there is an on-going battle between competing technologies for V2X communications (DSRC and C-V2X) that currently has the ITS industry reliving the days of VHS and Betamax. Only time will tell if this critical portion of wireless spectrum will remain available for smart transport and which standards will be utilized for V2X. What we do know is that while we are sitting around and waiting, we are losing lives.

4   Conclusion

Smart cities around the world are continually working to deploy technologies that enable applications that can save lives, improve mobility, and reduce environmental impacts. Despite the many technical and funding challenges facing smart transport today, the resolve of this industry will prevail over time and we will see increasingly smarter transportation systems in the future. As we've explored the various aspects involved in smart transport, even though the "What?" and "How?" aspects remain somewhat fuzzy, the "Why?" aspect is crystal clear.

References

1. World Health Organization. Global status report on road safety 2015. http://www.who.int/violence_injury_prevention/road_safety_status/2015/en/
2. INRIX Global Traffic Scorecard. http://inrix.com/scorecard/
3. Kenney J (2011) Dedicated short-range communications (DSRC) standards in the United States. Proc IEEE 99:1162. https://doi.org/10.1109/JPROC.2011.2132790
4. Kinney S (2018) What is C-V2X? RCR Wireless News
5. Hill S (2018) 4G vs. LTE: the differences explained. Digital Trends
6. US Dept. of Transportation, Intelligent Transportation Systems Joint Program Office. Applications for the environment: real-time information synthesis (AERIS) program. https://www.its.dot.gov/research_archives/aeris/index.htm
7. National Transportation Communications for ITS (NTCIP). https://www.ntcip.org/
8. RFC 1157. A simple network management protocol. https://tools.ietf.org/html/rfc1157
9. W3C (2007) SOAP version 1.2 part 1: messaging framework. https://www.w3.org/TR/soap12/
10. W3C. Extensible markup language (XML). https://www.w3.org/XML/
11. Institute of Transportation Engineers. Traffic management data dictionary (TMDD) and message sets for external traffic management center communications (MS/ETMCC). https://www.ite.org/technical-resources/standards/tmdd/
12. National Academies of Science. Cybersecurity of traffic management systems (NCHRP 03-127). http://apps.trb.org/cmsfeed/TRBNetProjectDisplay.asp?ProjectID=4179

13. Institute of Electrical and Electronics Engineers (IEEE). https://www.ieee.org/
14. Society of Automotive Engineers (SAE). https://www.sae.org/
15. International Organization for Standardization (ISO). https://www.iso.org/home.html
16. National Operations Center of Excellence. Vehicle to infrastructure deployment coalition (V2I DC). https://transportationops.org/V2I/V2I-overview
17. The American Association of State Highway and Transportation Officials (AASHTO). https://www.transportation.org/

Smart City Automation, Securing the Future

Adam Cason and David Wierschem

Contents

A. Cason
Futurex, Bulverde, TX, USA

D. Wierschem (✉)
Texas State University, McCoy College of Business, San Marcos, TX, USA
e-mail: dw50@txstate.edu

© Springer Nature Switzerland AG 2020
S. McClellan (ed.), *Smart Cities in Application*,
https://doi.org/10.1007/978-3-030-19396-6_5

1   Introduction

The speed of evolution of today's cities is unparalleled in history. In 133 BC Rome, Italy was the first city to reach one million inhabitants. In 2016 there were 436 cities with at least one million people. The top 10 cities all had populations in excess of ten million people [1]. The growth and success of these cities requires an ever expanding assembly of roads, pipes, cables, and people within a frenzied structure of concurrent connectivity and communication.

A continuous progression of technological advances has resulted in an interconnected web of information and services working together to transform cities into smart cities. These smart cities as defined by Techopedia use information and communication technologies to enhance the quality of living for its citizens. In the process, smart cities become increasingly dependent upon the reliability and security of those communication systems.

Smart city communication systems integrate a wide range of services from emergency responders to utilities; from citizen support to regulatory enforcement. The scope of information and integration continues to expand and the need for increased protection and security is compounded.

One example of the impact these technologies are having and the associated security tools used to protect them is represented by the continued progression of automotive technology.

In this chapter we will explore the opportunities that are quickly becoming reality in today's smart city automotive transportation arena. We will also look at the complexity and susceptibility of the underlying data systems that those opportunities rely upon. Finally we will discuss the security measures available to ensure the safety and reliability of those systems as just one component of the smart city.

2   Opportunities

The advancement in a variety of technologies that are being applied to transportation has occurred at a rapid pace. Foundational technologies such as global positioning systems, wireless communications, and artificial intelligence are enabling new products and services across a wide array of transportation areas. Examples include high speed rail, hyper loops, autonomous vehicles, and even drone delivery. Each of these advances have the potential to be a disruptive technology and they are all dependent upon integrated systems that share information.

The development of the automobile and their associated roads, stoplights, and other control technology resulted in an industry sector that accounts for $471.9 billion or 2.4% of the U.S. Gross Domestic Product (GDP) [2]. Today's new innovations are wide-ranging and include technologies such as autonomous vehicles, satellite-based tracking and guidance (GPS), smart roads, and magnetically levitating trains.

2.1   In Vehicle

The automobile of today is a far cry from just a few years ago. Today's vehicles include a myriad of electronic enhancements for both safety and pleasure. The first electronics were part of the operations of the vehicle. Ford introduced a computer controlled anti-skid system in 1969. General Motors introduced a computer controlled transmission in 1971. Cadillac introduced a computer controlled trip computer powered by a Motorola microprocessor in 1978 [3].

These advances were just the beginning. Today there are over 50 computer systems dedicated for monitoring and/or controlling everything from ride handling, to on-board entertainment and communication systems [3]. These systems are briefly diagrammed in Fig. 1.

Additional systems are being incrementally added that all lead to and support autonomous operation. These operational systems include night vision with pedestrian detection; automatic high-beam control; parental control; and GPS vehicle tracking. Operational systems are being augmented by passenger comfort systems such as in-vehicle internet access and entertainment systems supporting audio and video. As vehicles evolve to incorporate autonomous driving modes, passenger comfort and enjoyment opportunities will further increase.

2.2   Vehicle to Vehicle

More than six million crashes occur on U.S. roads alone every year, and more than 30,000 of those are fatal [4]. It has been estimated that adoption of autonomous vehicles will reduce vehicle related deaths by 90% [5]. Technologies driving this reduction include: autonomous driving; pilot assist with automatic lane positioning, pedestrian detection, and collision detection; and rear-mounted radar. Cars will talk to each other, automatically transmitting data such as speed, position, and direction, and they will take corrective action to prevent an imminent crash.

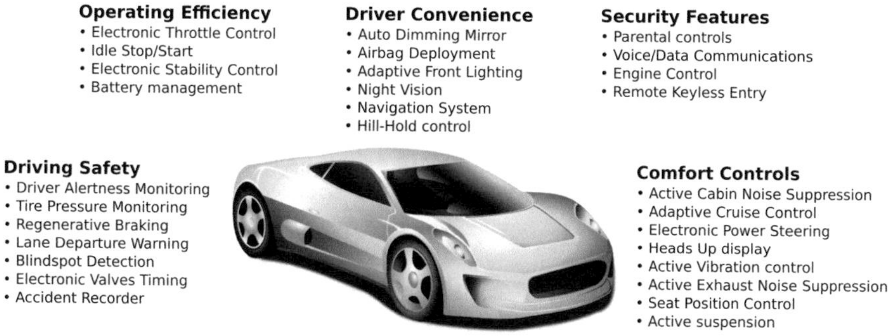

Operating Efficiency
- Electronic Throttle Control
- Idle Stop/Start
- Electronic Stability Control
- Battery management

Driver Convenience
- Auto Dimming Mirror
- Airbag Deployment
- Adaptive Front Lighting
- Night Vision
- Navigation System
- Hill-Hold control

Security Features
- Parental controls
- Voice/Data Communications
- Engine Control
- Remote Keyless Entry

Driving Safety
- Driver Alertness Monitoring
- Tire Pressure Monitoring
- Regenerative Braking
- Lane Departure Warning
- Blindspot Detection
- Electronic Valves Timing
- Accident Recorder

Comfort Controls
- Active Cabin Noise Suppression
- Adaptive Cruise Control
- Electronic Power Steering
- Heads Up display
- Active Vibration control
- Active Exhaust Noise Suppression
- Seat Position Control
- Active suspension

Fig. 1 Common computer systems in the modern vehicle [6]

2.3   Vehicle to Infrastructure

Not only are vehicles being made more intelligent, their operating environment is also being filled with intelligence. Examples of supporting infrastructure intelligence include: motion sensors that light up sections of roads only when a vehicle is in its zone; embedded magnetic fields that can charge electric vehicles while they are driving; road targeted police drones; weather and traffic detection; intelligent networked highways monitoring traffic flow, congestion priced toll charges, and the ability to coordinate driving speeds with traffic light patterns to maximize fuel economy and speed.

These technologies all work together to reduce collisions and fatalities as well as improve traffic flow and reduce fuel consumption.

2.4   Vehicle to Services

As these technologies are introduced, vehicle usage patterns will adjust and associated vehicle services will adapt to meet new demands and opportunities. Services that are already being explored include augmented reality for vehicle maintenance. More impacting is the movement of users away from traditional vehicle ownership towards greater use of shared mobility. Shared-ride services such as Uber and Lyft are enabling individuals to get where they want to be without the necessity or costs of ownership. This evolution will only continue as fleets of autonomous vehicles take to the streets.

Support services are also adapting. Patents have already been applied for a driverless police car. Microsoft is using Hexagon Safety & Infrastructure technology to improve and install first responder safety and efficiency into a smart city patrol car [7]. These combined technologies demonstrate how cloud-based and connected solutions can make cities safer and police more productive, resulting in an overall view of the smart city as a service.

2.5   Vehicle to City

As these new technologies continue to advance the very essence of the vehicle-to-city relationship will change. The evolution of autonomous vehicles will allow city dwellers to switch from owning a vehicle to partnering with car service providers such as Uber or Lyft. These companies will own and operate the cars, take care of licensing and maintenance, and for a fee provide automated transportation of humans to their destinations. They will also be able to provide courier/delivery services for business. Such a transition would result in reducing the number of vehicles on the roads, a reduction in space requirements for parking, and an associated reduction in carbon emissions.

3   A Definition of Trust

Smart city managers, service providers, and inhabitants all have lifestyle expectations living and working where they do. As exhibited with the transportation example, vehicles alone have created very high offerings and subsequent expectations that must be met. Such opportunistic efforts span throughout the whole transportation area as well as all the other services provided by and within the smart city.

The viability and success of all of these opportunistic efforts require a complex integration of hardware and software technologies that are designed and implemented with a focus of trust. Continuing with our vehicle example: a trust by the end users (drivers and passengers), of the providers (company designers and implementers) as well as the government (regulators and enforcers).

Trust is defined by Merriam-Webster as:

- assured reliance on the character, ability, strength, or truth of someone or something
- dependence on something future or contingent; hope
- a property interest held by one person for the benefit of another

Applied within the context of this chapter, trust can be defined as:

the inherent expectation of a person to experience safe, secure, and efficient interactions with, and transport by, a vehicle.

The trust expectation is unique to the type of user interaction/expectation and its scope.

3.1   In Vehicle

Within a vehicle, users have expectations for safety and comfort. Relative to technological systems users expect, at the most basic level, that they will work. And not only that they will work, but that they will work for the duration of their trip as well as the duration of their interaction with the vehicle. In the case of a rental or rideshare it will last the duration of the trip. If the vehicle is owned, then the expectation is for the duration of ownership.

Examples of trust expectations range from the simple: the radio, air bags, and temperature control to the more complex: Internet and/or phone access, cruise control, and entertainment systems.

A robust, Public Key Infrastructure (PKI) based security foundation can enable functionality like driver profiles. Imagine getting into a car and having your profile (radio stations, media library, seat position, navigation system favorite destinations, etc.) automatically loaded. When customer preference data is aggregated across an entire fleet of deployed vehicles, it also becomes a valuable dataset for auto manufacturers and their third-party partners.

At the same time, this dataset becomes a treasure trove of information for identity thieves. Just as the advent of big data has prompted security concerns among companies collecting personally identifiable information, so too will auto manufacturers be responsible for considering these risk factors. Often, information which may not be considered sensitive on an individual level can be deemed so when correlated with other data sets. Manufacturers must consider how this data is protected, both at the point of capture inside the vehicle and in their own data warehouses.

Surveillance, both by law enforcement and private entities, is also impacted. It's accepted today that where someone travels in their car cannot be hidden. Travel through toll booths, for example. That mindset would certainly change if this information became readily available electronically to anyone with even a casual interest due to insecure data protection mechanisms. The US Supreme Court has determined that GPS tracking of vehicles is protected by the Fourth Amendment [8], but criminal enterprises do not follow judicial precedent.

Authentication has become another major issue. Historically, to steal a car you needed physical access. This was accomplished by stealing the key, or breaking a window and then hot wiring the ignition. The introduction of chipped keys has made it more difficult. Authentication was associated with possession of the key. However, with the evolution of keyless start, authentication has taken a step back. Hackers are now able to trick a vehicle into thinking that its entry fob is nearby, thereby allowing entry and the ability to start the vehicle.

Today there are numerous methods of authentication, many of which can be handled using the vehicle's software systems rather than requiring additional hardware such as a key. One approach is to use a multi-factor authentication system, where a combination of attributes and authentication mechanisms is used to confirm driver identity. This could include biometric data such as a facial image from an in-vehicle camera or "voice print" authentication through existing microphones, or data analytics that use the individual driver's typical patterns to determine whether a request is legitimate.

An intelligent system for authentication can dramatically increase security without imposing an unnecessary burden on the driver. For example, a vehicle may learn that its owner typically uses their car to commute to and from work on weekdays. Under normal circumstances, starting the vehicle will only require the presence of a key fob. But if the vehicle receives an unusually timed start command, perhaps in the middle of the night, it would treat it as suspicious and automatically require a second or third factor of authentication such as the driver speaking their name. This does, however, raise questions about how overrides may be put in for legitimate needs such as medical emergencies.

Electronic payment systems are another important issue to consider. At the 2017 consumer electronics show, Honda revealed a prototype in-vehicle payments system for parking and fuel [9] With a little imagination this could easily be expanded to any automotive dependent services such as drive through eating establishments, car washes, or drive through entertainment venues.

It's likely that initial deployments of in-vehicle payment systems will use existing connectivity between smartphones and vehicles to route payments. This method rides the rails of the existing mobile payments infrastructure, which has well-established security standards and regulations. This includes cryptographic technology for card issuance and payment validation, typically deployed either on-premises or in the cloud, with Futurex's Excrypt Plus hardware security module or VirtuCrypt's cloud-based Crypto-as-a-Service being examples of those, respectively [10].

3.2 Vehicle to Vehicle

Multi-vehicle expectations deal primarily with safety concerns. Users expect these to not only work but to be invisible and unnoticeable. Automatic lane positioning should occur without notice of the passengers. Similarly, once globally installed, the necessity for emergency stopping and/or redirection to avoid obstacles should be greatly diminished.

From a security perspective, authenticity and integrity of vehicle-to-vehicle messages is much more important than data confidentiality. It could be argued that for the sake of safety and interoperability, V2V communication should not be encrypted at all. What's most important is that a vehicle knows when a legitimate message is being sent rather than a false one from a malicious actor.

Creation of an open standard for vehicle-to-vehicle messaging is critical for industry-wide adoption. While many aspects of transportation and feature implementation are proprietary to individual manufacturers, vehicle-to-vehicle communication must be interoperable.

How do you prevent false vehicle-to-vehicle messages from being sent? Beyond basic concerns about hackers creating false traffic jams, you could have actual accidents caused as a result of this. The answer to that is establishing a core of security within the vehicle and its components, where every vehicle is digitally signed under a common (likely manufacturer-specific, or by a government regulatory body) root certificate, which enables PKI-based mutual authentication and encryption of messages between vehicles.

3.3 Vehicle to Infrastructure

As vehicles become more connected to each other and to their environment the span of trust expectation widens.

- One area that has been highly controversial is the usage of ticketing cameras to monitor speeding, stopping at red lights, and more recently the illegal passing of stopped school busses. Ensuring that the driver of the car is the one awarded the ticket versus the owner of the car has caused many policing districts to reevaluate the usage of such technology.

- Another common area of interaction is with toll roads. As more states and municipalities view toll roads as ways to alleviate the financial burdens of highway costs, interaction with technology has increased. Increasingly, human-collected tolls are being replaced by unmanned and/or electronic collection systems. The use of tags mounted onto dashboards with monthly billing is becoming the norm requiring shared communication of license plate data between governmental agencies.
- Security of personal information by the government or service provider. The wireless, real time communications between the vehicle and centralized databases creates the opportunity for a variety of man-in-the-middle and other eavesdropping attacks.

3.4   Vehicle to Services

On the horizon, as vehicles are infused with technology and autonomous capabilities, the need for patches and upgrades will become a necessity. Manufacturer-specific data communication with the vehicle will be required to meet these demands. The ability for the manufacturer to pull data and monitor vehicle maintenance based on driving patters will result in new services being provided that span geographies. An example may be that monitoring devices notice a catastrophic break is imminent and therefore contact the driver and direct them to the nearest dealership for emergency repairs.

Such support service requirements will also provide new opportunities for services. Examples such as General Motors' On-Star service [11] which provides drivers with real time assistance for both safety as well as convenience.

Just as in-vehicle information about consumers should be considered sensitive, so should vehicle-to-service data. Aggregate repair and component failure data can be a rich target for industrial espionage.

3.5   Vehicle to City

The integration of technology into vehicles will result in the requirement for new regulations and controls. As new services expand offerings and replace existing businesses economic shifts in employment and revenue streams will impact the city and its inhabitants. Changes in emergency response protocols and penalty guidelines will need to be created. How do you give a ticket to an autonomous vehicle that is delivering a pizza?

Long-term city planning will need to be adjusted to accommodate for shifts in vehicle usage. Will fewer/more parking spaces be needed? Do electrical charging stations need to be provided?

Reliability of all the aforementioned services becomes increasingly important as the vehicles and associated services become more reliant on the communication infrastructure and the information it carries. Design issues such as redundancy and fail safe mode must be incorporated into all aspects of implementation.

4 Security Solutions

For smart city managers the expectations of trust require steps be taken to protect and ensure that trust. In many cases trust simply requires providers to execute and keep their promises. However, it also requires providers to implement processes and infrastructure to protect their processes and services from external threats. Within the confines of a smart city such efforts must be afforded by all participants. City inhabitants, or end users, are required to exert due diligence in their decisions to the extent they have control. However, providers and government have higher expectations of trust that must be met. Such expectations result in the necessity to implement more extensive security features that expand outside of specific situations such as "in-vehicle" and cover across a broad spectrum of instances and applications such as the transportation segment or even communication infrastructure in general.

4.1 Definition of Security

When architecting and deploying an information security system for distributed endpoints, regardless whether it's for smart city elements, automobiles, or Internet of Things devices in general, key principles must be followed. Many of these principles and best practices are well-established in the field of information security, but new challenges are also emerging that must be addressed.

At the heart of any information security infrastructure, a three-step process must be used to ensure information is secure as it's generated, exchanged, used, and stored. These items apply for data at rest (i.e., storage), data in motion (being transferred from one entity to another, also referred to as data in flight or data in transit), and data in use (data currently being processed in a temporary location such as a CPU, RAM, etc.):

- Authentication of the message's source and recipient, to make sure the sender and receiver, whether human or device, can trust one another.
- Integrity of the message, to make sure it has not been altered without knowledge during the course of transmission, storage, or use.
- Confidentiality of the message, using encryption to prevent unauthorized parties from viewing sensitive information.

With a solid information security foundation that relies on proven principles and is agile enough to incorporate new techniques, a constant umbrella of protection can be created to guard against threats such as network intruders, service disruptors, unknown service alterations, and trusted agent masquerades.

For our focus area of transportation infrastructure in particular, the threat of denial of service must be addressed. For smart vehicles, availability of services is critical. While much of security is focused on keeping information confidential, or verifying sender/recipient authenticity, or that nobody has tampered with the message, it's vital to make sure the infrastructure itself remains accessible and has plans for redundancy in the event of an attack or disaster.

If transportation systems are disrupted, vast portions of city infrastructure can become inaccessible to citizens. When automated vehicles or more widely used mass transit systems come into play, the risk of large-scale disruption grows immeasurably.

4.2   Establishing the Foundation of Trust with Encryption

Encryption is the most critical security element in protecting the confidentiality, authenticity, and integrity of data. Principles of strong encryption are well-known and have been integrated into security infrastructure for decades. By and large, it's not complicated, but it is processor-intensive. Modern CPU architectures are making this less of a problem, and many organizations exist worldwide with a focus on developing chipsets specifically optimized for cryptographic processing.

With the advent of large IoT and smart vehicle deployments, the implementation and operational burden shifts from encryption to key management. More devices and services than ever are generating data, and each of those devices require their own cryptographic keys and certificates. Managing the lifecycle of those keys and certificates can be incredibly complex.

Implementing a robust, scalable, highly available key and certificate management infrastructure is one of the most complex, and most important, challenges for smart city IT security managers. Management of the key and certificate lifecycle is typically performed using hardware security modules, such as Futurex's KMES Series 3 [12], that are certified to rigorous regulatory standards such as FIPS 140-2 Level 3 [13] and support a wide range of cryptographic algorithms, key types, and interface mechanisms.

Although outside the scope of this chapter, understanding the following foundational concepts is vital to building a working knowledge of the security requirements of smart vehicles.

- Encryption for user and device authentication

  - Key and certificate (PKI) management

- Key management lifecycle

  - Creation
  - Distribution
  - Storage
  - Deletion/revocation
  - Tracking/auditing

- Internet of Things (IoT) device authentication with PKI certificates

  - Device signing
  - Certificate authorities

- The role of mutual authentication in protecting smart city services/devices/data

4.3 Public Key Infrastructure: A High-Level Overview

Secure data is constantly being sent over public networks like the Internet. How, then, can users be sure their sensitive data will be protected and will only be accessed by the intended party? Any compromise or theft of sensitive data could mean millions of dollars in damages, or even the downfall of a company. In order to protect this information, users can encrypt their data, making it unreadable to anyone but the intended party. A public key infrastructure, or PKI, allows users to privately exchange sensitive data over public spaces by encrypting the data with a public and private cryptographic key pair that is created and shared through a trusted device, allowing both parties to be confident their sensitive data is protected.

4.3.1 What Is PKI?

A Public Key Infrastructure, which works by using asymmetric encryption, allows users to securely transmit sensitive data over insecure public spaces such as the Internet. By using PKI, this data is both encrypted and authenticated, enabling the recipient to be assured of the confidentiality and integrity of the message. Figure 2 diagrams the PKI encryption and key distribution process.

Public key infrastructures use public and private key pairs that are generated and distributed by a trusted device known as a certificate authority. Certificate authorities, which are often validated by third-party auditors, are used to generate digital certificates and assign them to the electronic devices that make up the PKI.

A certificate is made up of two parts: a public key and a private key. The public key is used to encrypt data and the private key is used to decrypt it. Public keys can be widely distributed without fear of compromise because public keys cannot decrypt data. Only the private key, which must be carefully protected, can successfully decrypt the message. When a single party provides their public key to another,

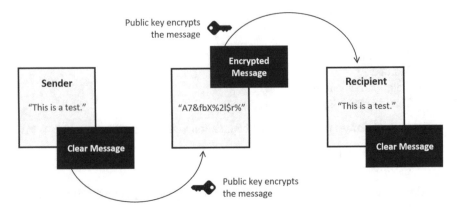

Fig. 2 Example of public key encryption

it enables a "one-way" channel to transmit data. When both parties exchange public keys, they are able to send trusted communication back and forth.

When all parties in a PKI are authenticated using the same certificate authority, a circle of trust is established. The entire certificate management lifecycle can be performed by the same certificate authority: introducing new trusted parties, revoking certificates from unauthorized parties, storing certificates, and periodically issuing new credentials upon their expiration.

4.3.2 How Does PKI Work?

Parties wishing to confidentially send data to one another begin by exchanging public keys. Using the receiver's public key, the sender encrypts their message and sends it to the receiver. Only the receiver's private key can decrypt the message. The two keys are related mathematically, but the private key cannot be derived from the public key, so there is no chance of compromise by sharing it. The private key, however, should never be shared under any circumstance.

These certificates are also used for signing and verifying secure data to ensure message integrity. In these cases, the private key is used to sign the message, which generates a certificate that contains information about the key, including the key owner's name. This certificate is sent along with the signed data to the receiver. The public key, which is available to all users, is used to verify the received data, ensuring that it came from a trusted source and was not forged.

4.4 Mutual Authentication: A High-Level Overview

When exchanging sensitive data electronically, users want to ensure their data is protected from compromise. Doing so requires confidence that the devices with which they exchange information are secure and protected. They must be able to verify that their information is sent to the intended destination and that intruders cannot gain access to this information in transit.

The solution to this issue is authentication. In the data security world, authentication occurs when users must prove their identity to log onto a computer, network, or other secured area. This can be accomplished in several different ways, the most common being entering a user ID and password. Devices that share information with other devices must also verify one another through what is called mutual authentication.

Mutual authentication is the process by which devices communicate with each other securely, guaranteeing the authenticity of the information being transmitted, preventing attackers from tampering with the data, and ensuring that data is not stolen or sent to an unauthorized device.

Similar to how people might present driver's licenses to identify themselves, devices verify one another's digital certificates to authenticate other devices. Digital certificates contain identifying information such as the name or address of the owner of the device. These certificates are issued by a trusted certificate authority device, which must be compliant with regulatory standards and housed within secure, independently audited environments.

Certificate authority servers manage the entire certification lifecycle: creating new certificates, monitoring certificate expiration dates, and revoking certificates. These certificates are used to form the basis of secure PKI, as they also contain asymmetric key pairs used for encrypting, decrypting, signing, and validating exchanged data.

4.5   Applications of PKI in the Automotive World

To secure information used by smart vehicles, a PKI-secured foundation of trust must be established.

This process begins on the manufacturing floor, when individual components are produced. This includes a wide range of items including engine control computers, safety systems, media players contained within the vehicles, and more. Each of these components must be injected with a private key and digitally signed by a trusted root certificate authority using established cryptographic techniques, allowing them to be verified as authentic and trusted components.

Once the entire vehicle is produced, it can be digitally assigned to its owner using the same techniques. Hypothetically using an application on their phone or computer, or even a key fob, the owner will generate a certificate signing request and send it to the manufacturer or dealer, who will sign it using their own trusted root certificate authority. Because the vehicle only trusts operators signed under that same root certificate authority, this process will bind the vehicle to its owner or operator. This can either be permanent or temporary in the case of shared ownership or rentals.

If necessary, the manufacturer or dealer can even remotely revoke the operator's authentication certificate. This can be used for tasks such as remotely disabling vehicles whose owners have not made payments. This also raises complex legal questions about the access law enforcement should have to this functionality as well. Should, for example, law enforcement have the ability to force a suspected criminal with an arrest warrant's autonomous vehicle to override their inputs and drive them to a police station?

4.6   Future-Proofing the Smart Transportation Infrastructure

Automobiles are designed for long-term use, and any deployment of smart vehicles must consider cryptographic agility. As stronger cryptographic algorithms and techniques are released, or security vulnerabilities are discovered, it must be possible to update them in the field.

When architecting the smart vehicle security infrastructure, designers must consider how they will support their customers far into the future. For this reason, as well as in support of vehicle-to-vehicle interoperability and message sharing, a standards body may emerge to allow certain industry-wide security burdens to be shared.

The security tools presented here are examples of the protective infrastructure required to maintain high-level trust required for not only vehicles or the transportation segment, but across the smart city. They are a necessary requirement to meet the trust expectations associated with ever increasing opportunities.

5   Concluding Remarks

The continued advancement and expansion of automation throughout the smart city will necessitate increasing efforts to ensure and protect the trust associated with increasing numbers of city inhabitants and associated products and services offered.

As presented, while implementation of automation takes place at specific products and services, i.e., our vehicle example, the protection and security infrastructure cuts across all industry segments and numerous implementations.

The variety of city components such as government services, emergency services, and utilities for example, each have corresponding unique automation security considerations, requirements, and solutions.

Ultimately, each component of a smart city has the potential to be a disruptive force in modern society, affecting work, recreation, and the day-to-day lives of virtually everyone. Where this disruption leads will depend on how city governments, businesses, and individuals work together to plan for the future while protecting the information and communications that fuel the change.

References

1. United Nations. The world's cities in 2016. Data book. http://www.un.org/en/development/desa/population/publications/pdf/urbanization/the_worlds_cities_in_2016_data_booklet.pdf. Date obtained 11/5/2-18
2. United States GDP From Transportation and Warehousing Summary. Trading Economics. https://tradingeconomics.com/united-states/gdp-from-transport. Date obtained 4/16/2018
3. Davis N. Automotive electronics: what are they, and how do they differ from "normal" electronics? In: Davis N. https://www.powerelectronicsnews.com/technology/automotive-electronics-what-are-they-and-how-do-they-differ-from-normal-electronics
4. Driver Knowledge. https://www.driverknowledge.com/car-accident-statistics/. Date obtained 7/12/2018
5. Clerkin B. Death by text message? Stats show how technology is killing us. DMV.ORG. https://www.dmv.org/articles/death-by-text-message-stats-show-how-technology-is-killing-us/. Date obtained 7/12/2018
6. ChipsEtc. http://www.chipsetc.com/computer-chips-inside-the-car.html
7. Hexagon. Hexagon safety & infrastructure's public safety app featured in the Microsoft patrol car. https://www.hexagonsafetyinfrastructure.com/news-releases/hexagon-safety-infrastructure-public-safety-app-featured-in-the-microsoft-patrol-car. Date obtained 7/12/2018
8. US Supreme Court, Torrey Dale Grady v. North Carolina, supremecourt.gov. https://www.supremecourt.gov/opinions/14pdf/14-593_o7jq.pdf. Date obtained 7/12/2018
9. American Honda Motor Co., Inc. Honda and visa demonstrate in-vehicle payments with Gilbarco and IPS Group at 2017 CES. https://www.prnewswire.com/news-releases/honda-and-visa-demonstrate-in-vehicle-payments-with-gilbarco-and-ips-group-at-2017-ces-300386576.html. Date obtained 7/15/2018
10. Futurex. Futurex launches next-generation financial HSM suite. https://www.futurex.com/news/futurex-launches-next-generation-financial-hsm-suite. Date obtained 7/18/2018
11. General Motors. Be safe out there. https://www.onstar.com/us/en/home/. Date obtained 7/19/2018
12. Futurex. KMES series 3. https://www.futurex.com/products/kmes-series-3. Date obtained 7/18/2018
13. FIPS PUB 140-2: Security requirements for cryptographic modules. National Institute of Standards and Technology. US Department of Commerce, 15 Nov 2001

Smart Cities Applications of Blockchain

Joe Moorman and Michael Stricklen

Contents

1 Overview of Blockchain Technology

Blockchains are a distributed and shared approach to recording digital transactions. Because blockchains are implemented without a central repository or management authority, they do not rely on trusted relationships between buyers, sellers, financial institutions, broker-dealers, or any legal entities. Rather, they are specially constructed to include tamper-evident and tamper-resistant mechanisms so that transactions are "incorruptible," or effectively impossible to manipulate after they are recorded. Additionally, blockchains are widely published (either publicly or across

J. Moorman (✉)
Myriad Genetics, Austin, TX, USA

M. Stricklen
EY Parthenon, Derry, NH, USA

© Springer Nature Switzerland AG 2020   101
S. McClellan (ed.), *Smart Cities in Application*,
https://doi.org/10.1007/978-3-030-19396-6_6

many nodes if private/restricted blockchains) and can therefore be copied in many locations continuously for redundancy and security. Bitcoin, first launched in 2009, utilized this technology to create the world's most prominent blockchain-based payment system.

1.1 Bitcoin's Meteoric Rise

What would you consider the most amazing machine of the past decade? Innovations such as the quantum computer, AlphaGo, or perhaps something more fundamental such as the memristor could be top of mind. Among these technologies, certainly the "blockchain hype machine" appears high on this list. Driven by investor exuberance and a new, novel approach to currency transfer, trading volumes of Bitcoin (BTC) rocketed up an order of magnitude or more, sending the price of 1 BTC from US$850 to 1100 to nearly US$20,000 over the course of 2017 [1]. That's some truly staggering growth by any measure.

Run the clock forward to February 2018, and one sees the price collapse to less than a third of its peak value [2], and by the end of this year the value of one BTC is well under US$4000 [3]. The hype machine had seemingly run out of fuel. Major news outlets had stated as far back as 2016, prior to the Bitcoin bubble, that cryptocurrency had run its course [4], a refrain heard again in 2018; however, these prognostications have no impact on the underlying viability of blockchain technology.

What was bad for speculative investors will ultimately be positive for legitimate uses of the technology. Large and small enterprises now view the technology with more equanimity and less mystification. The possibilities of deploying blockchain technology at scale are proven, and now organizations continue to find applications of blockchain that transcend the initial concept of a fully decentralized currency.

1.2 Implementation of a Blockchain

The fundamental architecture of Bitcoin was, and remains, a distributed ledger in the form of a blockchain [5]. A blockchain is implemented as a linked series of data nodes, aka blocks, where each block represents a single transaction on the system. The series of these nodes, the blockchain, contains the entire transaction history of a single distributed ledger. Compared to every previous payment system, the major distinguishing factor of blockchain is that it does not have a centralized single source of truth for transactions: every participant in the marketplace is equally entitled (by proportion of their computing power in verifying the blocks) to make a request to record the truth. For their trouble, and for providing this computing power, they receive a bounty, which users of the system support through the payment of transaction fees. This approach to validating transactions is known as a "proof-of-work" (PoW) system. The action of validating a transaction in a PoW blockchain is better known as mining, which we will shortly discuss in more detail.

Imagine an auction house with no auctioneer, with all participants allowed to make deals among themselves, and the result of every transaction recorded in a single ledger by majority vote of the participants. This record book could be considered a physical analog to a blockchain. Computers have (practically) infallible memory, and therefore do not forget or imagine numbers like humans do so in normal operation a computer is significantly more reliable than humans. Each participant in a blockchain is simply copying and appending to a linked list of immutable digital data which has potentially millions of redundant copies.

One may then ask the question: what would stop a bad actor from forging a transaction? Blockchain does not give any credence to trustworthiness, and brings no requirements to establish the identities of any participant, but in spite of these constraints there are built-in safeguards [6].

Any individual person can attempt to forge blocks, but they will fail unless they control computational resources in excess of 50% of the participants in the blockchain. Similarly, a computer can try to modify transactions in a block, but to actually encode this in the record when the "majority vote" is determined by the proof-of-work algorithm means that you must perform a majority of the work. Again, this requires the bad actor must control a majority of the computing power on the network. Within the Bitcoin network, cheating with brute force would require computing hardware capable of over 30 quintillion hashes per second as of 2018: thousands of times more powerful than the world's top 500 supercomputers combined. To run this setup would cost trillions of kWh per year, more than the power bill of entire countries. One can see that for blockchain networks which have reached a critical mass this PoW control is effective for blunting attacks; even for smaller emerging blockchains, there are techniques to prevent brute force attacks on the validity of the chain [7].

1.3 Distributed Proof

One characteristic of blockchain is that it thrives on a widely distributed base of users and operators. Participation is encouraged by offering all parties equal entitlement to record the true state of the ledger, and the incentive to join the network is in direct proportion to the cost of their computational time and the electricity to run the exchange. Although a large enough company, such as a multinational bank, can independently run a private blockchain on machines it controls, public blockchains enable truly distributed organizations with no central coordination to emerge.

The first advantage of public blockchains is to deliver the transfer of value (currency) at lower costs. With the value of "a central trusted authority" displaced, the previous example of a blockchain auction house would have no need to pay auctioneers; therefore, its personnel costs are lower [8]. When applied to payment clearinghouses or any industry with an auditable paper trail, blockchain will likewise reduce regulatory reporting costs.

A further advantage of distributed proof is security and redundancy [8]. When a blockchain attains critical mass, the bounty means that participants are clamoring to record blocks to the blockchain, and transactions are virtually impossible to erase or modify. If a large bank were hacked, the loss of a handful of servers could effectively wipe out accounts, whereas hijacking the blockchain requires either brute-forcing falsified transactions or hacking an infeasibly large number of network nodes.

2 Smart Contracts

The smart contract is the future of Bitcoin and Ethereum and other cryptocurrency platforms. If you start from the premise that a blockchain represents a reliable, immutable history of transactions, the next step is to leverage the transaction block to express programmatically encoded rules which embody complicated payment terms.

2.1 Beyond Pure Cryptocurrencies

Swapping Bitcoin or Ether faces difficulty in one question: why? Unfortunately, there is some cost to running the proof-of-work algorithm, Hashcash, that underpins Bitcoin—this transaction cost is a function of the number of people using the platform [9]. Thus higher popularity for a blockchain network translates into greater time required to complete a transaction (but you can also increase your transaction fee offered to get a faster slot). The average clearing time for a single Bitcoin block is roughly an hour, but it can sometimes be much longer [10], and while Ethereum adds new blocks in well under 30 s [11] and can scale to a national clearing payment system [12], there are no governments that accept any cryptocurrency as legal tender, and many businesses remain unconvinced to accept it. Companies that accepted Bitcoin in 2017, during the volatility of the early period of Bitcoin, stopped accepting it when the value of the currency dipped. In order to return value to the participants who agree to accept these costs and the risk due to price volatility of taking a position in a cryptocurrency, participating in such an exchange must provide some major advantages over forking over a pile of hard currency.

The advantage is in competing with existing payment networks that provide built-in proof of transactions, ensure the identity of parties in the transaction, and enable complicated legal contracts which take a good deal of time by humans to craft and execute. In these situations, the self-correcting nature of blockchain provides a compelling advantage in terms of transaction cost.

2.2 Future Applications

The application of smart contracts is effectively unlimited in the payment system world, because they can encode any detail of any contract: they can hold funds in escrow, specify multiple payment channels, require multiple signatures to execute, or any set of conditions that the buyer and seller could possibly imagine. The prop bet you place on the next sporting event, or your mortgage could be written as smart contracts! Any such contract would be just one more transaction on the blockchain, albeit a more complex than using Ethereum as a currency, say to buy a cup of coffee.

3 Concept for a Distributed Mobility Service Network

Here we present a concept of a blockchain-based mobility service network, which has the potential to disrupt existing providers such as Lyft, Uber, Didi, and hundreds of other smaller players. Using a blockchain-powered ridesharing system, drivers could offer the use of their cars as a disjointed group of service providers, either as a cooperative or in competition with one another.

3.1 Legacy Mobility Services

Billions of rides every year are booked on mobility service providers like Lyft, Uber, and DiDi [13, 14]. In cities with limited or nonexistent public transportation networks, they are relied upon by millions of Americans on an almost daily basis. All these services share a common feature—a centralized, proprietary database which logs all your trips, all of your riding habits, and uniquely ties this to you individually. As one may expect, this data can be monetized for targeted advertising, which may or may not concern you as an individual.

The centralized nature of this database brings little additional value to the service providers within the ridesharing network: network operators hide their proprietary information from both drivers and riders, unless forced to disclose these records through court orders. Moreover, each company maintains its own data store completely independently of all other network operations. Lyft has no visibility on traffic patterns in Uber-dominated areas, for example. Competition is a good thing, but with at most 2 or 3 private rideshare players for a given market, and with labor costs only likely to increase, there is little incentive to innovate on the current business model. If you like the current ride-sharing reality, expect more of the same for the short-to-mid term.

3.2   Impact of Autonomous Vehicles

There is a vast untapped market in monetizing the use of privately owned vehicles which sit idle for a majority of the day. The intra-city trips taken with these private autos greatly outnumber those on mobility service providers [14]. Currently there's an extremely good reason why they sit: their owners must choose between being a driver or holding a different job, and most people do not work as drivers. Extrapolating based on current trends, privately owned autonomous vehicles could soon be ubiquitous. Developments by Tesla's Auto Pilot were the first to achieve Level 2 autonomy [15], where a vehicle possesses the capability to operate while still under continuous oversight by a human operation. Volkswagen AG has made arguably more progress toward Level 3 autonomy [15], where driver oversight is still needed but only when requested; all normal operation is done by the vehicle. Google's Waymo division has taken the most ambitious approach, with LiDAR sensors on their vehicles and millions of miles of fully autonomous driving conducted safely [16]. The cost of the sensors highlights the progress made. LiDAR costs Waymo $150,000 per vehicle in early 2016 test models, but this fell 90% by 2017 [17], and continues to become cheaper in subsequent models. GM is also close to fielding real autonomous vehicles, showcasing its Cruise AV based on the Chevrolet Bolt, in January 2018. Deployment of that vehicle is pending NHTSA petition [18], but could happen as soon as 2019 with 2500 vehicles [19] on the road. Therefore we're left to ask the following question: How does blockchain disrupt the existing model of mobility service providers, and what are the costs and benefits?

3.3   Example: The "Ridecoin" Network

Building upon legacy ridesharing platforms' business model, upstart platforms using community-based transaction ledgers, built on blockchain technology, will enable entirely new competing networks. As an example, Ridecoin is a peer-to-peer ridesharing network which leverages blockchain in place of a centralized middleman to manage and track payments. The result is a lower-cost service which returns more value to motorists and riders, enables additional public transportation options, and yields valuable data to the communities they serve.

3.3.1   Concept and Benefits

Let's fast-forward 10 years to a medium-sized Smart City with perhaps 25,000 or so vehicles that are in full-time use by mobility service providers like Lyft and Uber, most of them fully autonomous. Let's further assume these incumbents fared well on their IPOs, and although they burned money in their early years, they have been profitable for several years now, and they were among the first to launch autonomous vehicles commercially for their own services. Let us also assume autonomous technology has found its way into the far-more-numerous privately owned vehicles, say there are 100,000 autonomous vehicles in the hands of private owners in this city.

What is the commercial potential of these vehicles? Individual rides can be bought and sold for cash, but that is impractical enough that most owners don't use their autonomous vehicles commercially. Uber or Lyft may try to defuse the threat of these vehicles through pricing pressures, but not to an extent where it jeopardizes their market position, so very little changes. Using a blockchain-powered ridesharing system, even a small portion of those 100,000 motorists monetizing the idle time of their vehicles could disrupt that market.

We could see the scenario playing out thusly: Using CarPlay or Android Auto, one could run a "Ridecoin" app that offered the use of your vehicle to others using the app. Transactions would be encoded on the blockchain with all ride details: expected duration, distance, pickup time, liability clauses on damages that the rider may make to the interior, the rider count, and the price of the ride. You as an individual could undercut Uber or Lyft on price as you don't have the fixed cost of their business. All transactions would be public, so your own Ridecoin endpoint would have visibility on the price of transactions conducted recently, in order to decide when it is profitable to accept rides.

If you left Ridecoin running on your car when you parked at work at 7:59 am, you might get a push notification on your iPhone stating that it found a fare right away at 8:00 am, to pick up another commuter. It may find fares around your city from 8:00 am to 9:51 am, and from that point see business tail off thus returning to your workplace garage and parking itself. When you return to drive home at the end of the day, you would have earned Ridecoin (RC) for the use of your car—which you could then use to pay for the use of other drivers' cars in the same manner. The monetary benefits to the vehicle owner are obvious, but others in our community will also see a benefit.

Public enterprises could be the vanguard of such blockchain-based ridesharing platforms [20]. A city government will someday conclude that maintaining its human-operated bus and minibus fleet is more expensive than acquiring a fleet of autonomous vehicles. These autonomous vehicles could rove freely or make the same stops as the human drivers, but they would need a means to identify when passengers were present at a stop who wished to board. This could be done with a blockchain which the city government developed and sold to the public, just as they would bus passes and MetroCards. The government could orient policy directly to help the needy by distributing the currency over this blockchain to those who qualified for such benefits.

Policymakers have another surprisingly difficult task of making sure that persons with disabilities in our communities have compatible transport for their wheelchairs and other equipment, where even subways and city buses may be too difficult to use. New York's Access-a-Ride program, for example, gives discounted cab rides or shuttle rides as a way to enhance the lives of those with disabilities. Using their own blockchain or whichever system was dominant in the region, the government could acquire the ride currency and give it directly to their disabled citizens, which would organically stimulate the supply of handicapped-accessible vehicles for all those who used the same blockchain to provide rides. This would build on existing efforts in many cities to use blockchain technology to give digital identities to the homeless who lack conventional identification, such as Blockchain for Change in New York [21].

Benefits for city planning would flow from the availability of the ride data on the blockchain. In a world where a very sizable quantity of rides is transacted on the blockchain, data collected and processed would give the government a freely obtained view into traffic patterns, which would let them more efficiently allocate funds for road improvements, and make more informed decisions about land development. Still each transaction in the blockchain could anonymize identities, tying accounts only to specific hidden keys, preempting many of the concerns privacy advocates have regarding governmental surveillance of citizens.

3.3.2 Costs and Challenges

Direct costs of Ridecoin would be borne collectively by motorists, as their vehicle batteries would be used to power the system. Clearly its power consumption would need to be designed to be supportable by EVs and not significantly detract from their useful range. Politicians, under pressure from constituents, may decide it is wasteful to have unoccupied autonomous cars seeking fares throughout their neighborhoods, and impose a tax that makes the use of one's own vehicle unprofitable unless operating at the scale of a mobility service provider like Uber, making the distributed approach less feasible.

Whether owned by existing mobile service providers or blockchain upstarts, autonomous vehicles will need to be subject to an adequate taxation and regulation framework by city governments to avoid chaos, as the driver previously was the most expensive part of any commercial vehicle. This may take the form of advanced "curb space" APIs that tax the space that vehicles sit at to pick up fares, which Las Vegas is currently planning [22].

Governments running their own municipal transit blockchains, as proposed here, will be shifting labor costs associated with managing payment systems and transaction fees, to a purely technological solution based on a blockchain ride currency. When autonomous vehicles are widely adopted, one could conclude that cities will be under pressure from taxpayers to replace drivers of trains and buses, but unions will strongly resist [23]. Even if it is inevitable that autonomous vehicles will be bought by governments, the choice between simply automating existing bus routes and buying smaller autonomous cars to make a larger number of personalized journeys requiring individual identification will be one that helps them decide if blockchain is a viable solution.

4 Concept for a Distributed Microgrid

Blockchain has enormous implications for the power grid of a smart city. With the current model of a "macrogrid," a small number of extremely large power companies have billions of dollars of capital and tend to dominate all power production and distribution in their region. We present a concept leveraging blockchain for a distributed "microgrid" of very small power generating nodes which individuals and small businesses can operate, to commercialize their renewable investments, buy green power directly from their communities, and help ease the peak power generation derived from fossil fuels.

4.1 Historical "Microgrids"

Compared to humans of the twentieth century, humans of antiquity and the medieval period would not have thought of a "power plant" as something geographically remote, nor on an industrially large scale. They would have thought of a windmill or water wheels, which were the "power plants" of their time [24], and these were always geographically close to the area which they served. Throughout most of human history, virtually all examples of converting mechanical energy from the environment into useful work would be enterprises on the scale of a single village or neighborhood.

4.2 Dominance of the Macrogrid

Throughout the nineteenth century, stationary steam-powered engines, and later electrical power plants saw nonrenewable resources such as coal, oil, and natural gas supplant muscle as the dominant fuel source. Electricity was distributed through large grids that crossed states or nations. Large capital investments were required to develop plants of sufficient size to efficiently produce power, which remains true for nonrenewable energy, and has made macrogrids dominant well into the twenty-first century.

4.2.1 Fossil Fuels and Location of Plants

When the first electrical power plants were developed in the 1880s, power was distributed across the grid using direct current (DC). This had extreme transmission losses, and therefore required that plants be close (ideally within 1–2 miles) to the end users. The stubborn need for collocation with a DC plant was relieved with the introduction of alternating current (AC) at the end of the nineteenth century.

AC grids efficiently operated at much higher voltages and considerably reduced transmission losses; it spelled the obsolescence of DC power plants within two decades [25].

Even though the greater transmission efficiency allowed favorable land prices to affect moving power plants further from cities, a co-equal accelerant was the sheer dirtiness of fossil fuel-burning plants. The soot and pollutants that coal-fired plants generated precluded immediate development nearby, and wrought additional threats like acid rain. When generation plants were located far from cities, these were considered manageable drawbacks for the time.

Starting from this nexus and up until very recently, economies of scale and land costs still favored larger installations further from cities. Coal-fired generating stations would be the norm for the next 100 years, with hydroelectric to be added wherever geography favored it (typically not immediately near cities). Nuclear plants, for safety reasons, also needed to be physically distant from large population centers, and located adjacent to water sources with sufficient capacity to provide thermal controls.

4.2.2   Renewable Sources

Climate science experts have established a need to utilize renewable energy sources as rapidly as possible. If the goals of the Paris Agreement are unfulfilled and serious steps not taken toward carbon neutrality by 2050, it may be too late to avert catastrophic climate change [26].

Some major steps toward more environmentally friendly grids have already been taken. Cleaner-burning oil and natural gas power plants are significantly better for the environment in which they operate in terms of particulate and harmful pollutants, and they could be built closer to cities to reduce transmission losses, but they still emit too much carbon pollution. Among the available fossil fuel stocks, Coal is the worst offender at about 820 grams of CO_2 equivalent per kWh (gCO2e/kWH) for an average plant in the USA, while natural gas is better at 490 gCO2e/kWH [27]. For comparison, photovoltaics carbon emissions are less than a tenth of gas over their entire usable life.

Because it was easy for small organizations to take small steps toward carbon neutrality, we're familiar with renewable sources by now, as they have entered our daily lives. Renewables have achieved practicality because wind and solar can efficiently operate on a nearly infinite granularity of size: wind turbines as large as a skyscraper scale down to rooftop use for a business office or university lab, while photovoltaic panels can both stretch across miles of distant desert land or lay on the roof of a factory or home.

The major drawback of renewable sources is that they don't continuously produce. Where there is less wind or where windspeed is too high to safely operate, wind turbines produce less power or cease to produce it. When it is nighttime, PV produce no output; when cloudy, output will dip. This gap between peak demand and renewable production ability for a mostly green grid creates a characteristic

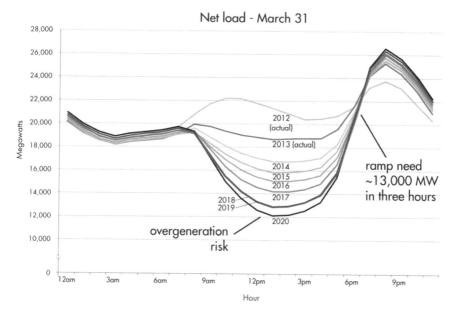

Fig. 1 Duck curve for California Independent System Operator (CAISO) for 1 day [28]

shape known as a duck curve (Fig. 1), for which there is a huge leap in the demand for power following the end of the business day, which runs the risk of wasteful overgeneration or keeping nonrenewable plants on standby.

There will remain some need for "baseload" power, which is always available in any weather conditions. It is probable that this is going to have a large fossil-fueled element for decades to come, but a significant portion of it can be offset with power storage solutions. In the meantime, during the sunny and windy daytime, the renewables need to overproduce enough that they fill enough storage to last through the night.

A macrogrid can most definitely use renewables with efficiency, but so too can businesses and individuals who install them for personal use [29]. Currently, there is a combination of both. The power companies try their best to cope with continuous requirements for baseload and make their grid as green as possible while still retaining profitability. Customers want inexpensive electricity and ascribe at least some value to whether it was green; many go as far as installing their own solar panels or wind turbines.

4.3 Microgrid Resurgence

The single big grid has limits on its flexibility [30]. Most power companies must keep their CO_2-producing plants running on part-time, because even if it their renewables would meet their daytime load, they'd run out of juice at nighttime.

Producing vast arrays of batteries for energy storage for no other reason than peak load is extremely expensive compared to firing up more plants that already exist. Customers who have a considerable investment in solar panels will find that they are overproducing for their own needs during the day, giving them the option of selling the power back to the grid. On the other hand, in the peak demand period just after working hours, renewables can only hang on until sunset and then drop off rapidly. Other dirtier plants must spin up to support this huge surge of power until late evening when it tapers off again.

4.3.1 Impact of Electric Vehicles

Facilitating the transition to decentralized microgrids, a massive increase in storage capacity will likely be precipitated by individuals. It will not be purely driven by a desire to monetize their storage, but will likely be induced by the purchase of an electric vehicle.

Across the USA at present, 70–100 million internal combustion cars might be part of the commute every day of the week. These vehicles produce CO_2 and other pollutants, but they neither help nor hurt the power grid's ability to run on green energy. As we continue to buy more and more EVs, if 100 million EVs go home empty and get charged immediately, they would tax the grid at its most vulnerable time.

Some have even used this as an argument against EVs, but this is spurious. A mostly fossil fuel-powered grid is still producing less CO_2 to charge an EV than an ICE vehicle would otherwise produce [31]. According to 2014 data, using an average regional macrogrid in the USA, which has a mixture of coal, gas, and oil plants to fill in peak demand, an ICE car would need to achieve better than 50 miles per gallon to be competitive, which is virtually non-existent in the USA. In the dirtiest regional grid in the USA, which is coal-heavy during peak, this can drop to 35 mpg (still higher than most ICE cars manage). But this is only true of some regions in the interior of the USA, totaling <17% of the US population. Meanwhile, across all grids, coal is being retired at a rapid pace and the grids are continuing to source increasing loads from green production methods. In the past 10 years, coal's share of electricity generation dropped from 50% to 30% in the USA [32], and it is poised to drop to about 20% by 2024 at currently predicted retirement rates [33].

Taking this concept further, the assumption that all EVs will be charging from 0% to 100% right after work is highly unrealistic. Assuming an average round-trip commute distance of 32 miles, with most workplaces able to supply L1 power [34], and mass-market EVs currently possessing more than 225 miles range, the nation's 100 million EVs will go home in varying states of charge—with some of them mostly empty and some of them mostly charged. In that fleet of vehicles is a cumulatively massive amount of energy storage which can push everyone through the peak demand period and prevent the need for tapping those dirty fossil fuel plants.

All of us who drive around in EVs would be putting into the hands of the grid designers some amount of storage to be intelligently utilized. So, designers simply need a means to incentivize motorists to sell power from their batteries during periods of peak demand, and recharge those batteries during the periods of peak renewable production.

4.3.2 Example: The "EnviroCoin" Network

To efficiently utilize all the distributed electrical storage of the future while reward-ing consumers for their renewable investments, and to allow power to enter the grid locally with less transmission loss, a concept for distributed exchange of power is presented. It will take the form of a microgrid blockchain, which would use smart contracts for power transactions, underpinned by a unit called "EnviroCoin" (EVC).

4.3.3 Concept and Benefits

Individuals and businesses who have surplus renewable energy will have a combi-nation of sources: PV cells on their rooftops, EVs sitting in the driveway or parking lot, particularly at work [34], and possibly even on-site storage such as Tesla's PowerWall or PowerPack. These entities will be able to acquire hardware which lets them sell power back to microgrids in the same form as macrogrids currently pur-chase it, but with great flexibility on a near-real-time basis. "Micro-providers" will gain this flexibility by utilizing integrated rules engines that permit transacting energy based on hourly prices, availability, weighting factor for renewable sources, weighting factor for resiliency, distance to source, or many other parameters [35]. Demand by consumers will imply the price; they will buy power on their local microgrid for EVC on a scale from cheapest at daytime to costliest at peak demand, to in-between in the dead of night. Where the microgrid lacks capacity or doesn't have a competitive price, the macrogrid can step in, but if it is profitable to compete with them, markets will react and small operators will add to their capacity.

Compared to macrogrids, the microgrid network will let customers buy electric-ity locally, reducing the transmission losses it took to receive it, and it will add redundancy by having more possible sources, as compared to a few power plants. It will also democratize the energy exchange process, making it possible for individu-als or small businesses to monetize their green investments, and make more trans-parent the costs of using electricity at more burdensome times of day.

Consider the stories of multiple users who interact with the microgrid:

- Motorist 0 uses an ICE car because she drives for a living as a courier, covering about 500 miles a day. Her home has PV panels installed, which exceed her day-time use. She finds that using the microgrid EVC platform, she can sell directly to locals for a higher price than the macrogrid will buy from her. Customers of her power know that it is produced with less transmission loss due to locality, and that it is 100% renewable.
- Motorist 1 has an EV but no charger at work. He cannot charge anywhere but his own home, and he has a long commute that almost empties his EV's battery

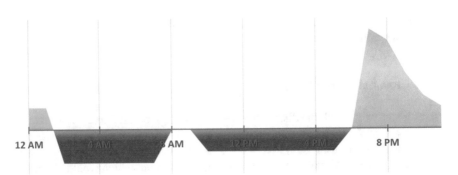

Fig. 2 Motorist 2's net electricity revenue, weighted by cost when acquired and price when sold

daily. For convenience, he splurges on a fast charger which he uses right after he gets home. This is right at the period where the microgrid generating capacity is weakest, so he is forced to buy some dirty power to cover the shortfall. Yet, recall that even the most careless EV user is likely still reducing CO_2 compared to the very best combustion engines [30].

- Motorist 2 has an EV and can charge at work. She charges up to 100% by quitting time, has 80% battery life by the time she gets home, and then lets her EV pump power back into the grid, acquiring EVC until it goes all the way down to 20% in the middle of the night, then charges it back up to 40% before the next morning, which is more than enough to get work where she can repeat the process. This is both the best possible use of renewable power resources, and a profitable arrangement for her, since she gets cheap green power during the day, which she sells at a higher price during peak periods. See Fig. 2.

4.3.4 Costs and Challenges

Costs of running the EVC platform for local energy exchange comes in the form of the energy which is used to run the blockchain. It will be borne as a tax on the cost of doing business, itself a transaction cost.

Throughout the transition, the cost of investing in microgrid technology may detract from corresponding upgrades to the macrogrid. A region with a strong prevalence of affluent communities investing in their own microgrids in an unregulated manner may imply that other communities in the area are underserved in their own power needs. Smart cities may need to take appropriate steps to regulate access to microgrid resources if the legacy utilities can no longer serve all customers well, perhaps subsidizing the cost of equipment that lets customers purchase from the microgrid.

Intermediate-term costs for power consumers may be higher as the distributed structure of the microgrids is developed. Large, established utilities must keep their large-scale generating power available in parallel during the transition to microgrids, as weather events can imperil the ability of microgrid users to generate local solar or wind power before advances in energy storage can support a full migration from centralized generation. High availability will be required to inject energy into the grid by large utilities during the transition, until microgrids are strong enough to meet the needs of everyone in their communities.

5 Conclusion

Blockchain will dovetail with many daily-life use cases of Smart Cities. By making publicly available a means of exchanging value with greater security and transparency, blockchain can support efficient, decentralized solutions to mobility services and power grids. Autonomous and electrically powered privately owned vehicles are key developments that will catalyze the development of these blockchains. The cost of the transition will first be borne by individuals and entrepreneurs who seek to profit from an initial stake, but as the blockchain userbase grows, society will need to address the needs of those who are less able to access their accustomed services during the transition, including the indigent and the elderly. Ultimately, as the costs of the technology to use the smart city blockchains fall with mass adoption, a complete transition to decentralized power distribution and decentralized mobility services could provide a maximally efficient solution for our society.

References

1. Morris D (2017) Bitcoin hits a new record high, but stops short of $20,000. Fortune [Online]. http://fortune.com/2017/12/17/bitcoin-record-high-short-of-20000/. Accessed 12 Jan 2019
2. Rooney K (2019) Bitcoin falls 7 percent to below $8,000 after Twitter's ban on cryptocurrency ads. CNBC [Online]. https://www.cnbc.com/2018/03/26/bitcoin-falls-7-percent-to-below-8000-after-twitter-bans-cryptocurrency-ads.html. Accessed 12 Jan 2019
3. Josiah W (2019) Bitcoin mining difficulty just saw its second-largest drop in history. CCN [Online]. https://www.ccn.com/bitcoin-mining-difficulty-just-saw-its-second-largest-drop-in-history/. Accessed 02 Jan 2019
4. Wadhwa V (2016) R.I.P., bitcoin. It's time to move on. Washington Post [Online]. https://www.washingtonpost.com/news/innovations/wp/2016/01/19/r-i-p-bitcoin-its-time-to-move-on. Accessed 02 Jan 2019
5. Peck ME; IEEE Spectrum Staff (2015) The bitcoin blockchain explained. IEEE Spectrum [Online]. https://spectrum.ieee.org/video/computing/networks/video-the-bitcoin-blockchain-explained. Accessed 12 Jan 2019
6. Peck ME (2017) How blockchains work. IEEE Spectrum
7. Chen L, Xu L, Gao Z, Lu Y, Shi W (2018) Protecting early stage proof-of-work based public blockchain. In: 2018 48th annual IEEE/IFIP international conference on dependable systems and networks workshops (DSN-W), Luxembourg, Luxembourg. IEEE, pp 122–127. https://doi.org/10.1109/DSN-W.2018.00050
8. Cognizant (2017) Demystifying blockchain [Online]. https://www.cognizant.com/whitepapers/demystifying-blockchain-codex2199.pdf. Accessed 12 Jan 2019
9. Fairley P (2017) The ridiculous amount of energy it takes to run bitcoin. IEEE Spectrum [Online]. https://spectrum.ieee.org/energy/policy/the-ridiculous-amount-of-energy-it-takes-to-run-bitcoin. Accessed 11 Jan 2019

10. Buchko S (2017) How long do bitcoin transactions take? CoinCentral [Online]. https://coin-central.com/how-long-do-bitcoin-transfers-take. Accessed 12 Jan 2019
11. Bitinfocharts (2019) Ethereum block time historical chart [Online]. https://bitinfocharts.com/comparison/ethereum-confirmationtime.html. Accessed 17 Jan 2019
12. Creer D, Crook R, Hornsby M et al (2016) Proving Ethereum for the clearing use case. Royal Bank of Scotland [Online]. https://cdn.relayto.com/media/files/lBFCJiHESx64mWhBAlzs_emeraldTechnicalPaper.pdf. Accessed 19 Jan 2019
13. Uber Technologies Inc (2019) Uber company info [Online]. https://www.uber.com/newsroom/company-info. Accessed 19 Jan 2019
14. O'Dea J (2017) How many rides do Lyft and Uber give per day? New data help cities plan for the future. Union of Concerned Scientists [Online]. https://blog.ucsusa.org/jimmy-odea/how-many-rides-do-lyft-and-uber-give-per-day-new-data-help-cities-plan-for-the-future. Accessed 19 Jan 2019
15. FundSpec.IO (2017) Audi beats Tesla (And GM) to level 3 autonomy. Seeking Alpha [Online]. https://seekingalpha.com/article/4087480-audi-beats-tesla-gm-level-3-autonomy. Accessed 19 Jan 2019
16. Ohnsman A (2018) Waymo is millions of miles ahead in robot car tests; does it need a billion more? Forbes [Online]. https://www.forbes.com/sites/alanohnsman/2018/03/02/waymo-is-millions-of-miles-ahead-in-robot-car-tests-does-it-need-a-billion-more. Accessed 19 Jan 2019
17. Naughton K, Bergen M (2017) Alphabet's Waymo cuts cost of key self-driving sensor by 90%. Bloomberg [Online]. https://www.bloomberg.com/news/articles/2017-01-08/alphabet-s-waymo-cuts-cost-of-key-self-driving-sensor-by-90. Accessed 19 Jan 2019
18. United States National Highway Traffic Safety Administration (2018) Petitions for exemption received during this administration [Online]. https://www.nhtsa.gov/laws-regulations/petitions-nhtsa. Accessed 19 Jan 2019
19. Wayland M (2018) GM ride-hailing fleet would ditch steering wheel, pedals in 2019. Automotive News [Online]. https://www.autonews.com/article/20180112/MOBILITY/180119919/gm-ride-hailing-fleet-would-ditch-steering-wheel-pedals-in-2019. Accessed: 19-Jan-2019
20. Rainwater B (2018) For smart cities, blockchain technology opens new possibilities. National League of Cities [Online]. https://citiesspeak.org/2018/07/09/for-smart-cities-blockchain-technology-opens-new-possibilities. Accessed: 19 Jan 2019
21. Schiller B (2017) This new blockchain project gives homeless New Yorkers a digital identity. Fast Company [Online]. https://www.fastcompany.com/40500978/this-new-blockchain-project-gives-homeless-new-yorkers-a-digital-identity. Accessed 19 Jan 2019
22. Lattin P (2018) How Vegas will monetize autonomous vehicles. Pace Vegas [Online]. https://pacevegas.com/2018/10/vegas-will-monetize-autonomous-vehicles/. Accessed 19 Jan 2019
23. Metro Magazine (2018) Transportation workers form coalition to stop driverless buses in Ohio. Metro [Online]. http://www.metro-magazine.com/management-operations/news/731336/transportation-workers-form-coalition-to-stop-driverless-buses-in-ohio. Accessed 19 Jan 2019
24. Langdon J (2008) The windmill: a medieval 'steam engine'? Presented at 6th technology and human capital formation in the east and west S.R. Epstein memorial conference, London School of Economics, London, UK
25. United States Department of Energy (2018) The war of the currents: AC vs DC power. energy. gov [Online]. https://www.energy.gov/articles/war-currents-ac-vs-dc-power. Accessed 19 Jan 2019
26. Masson-Delmotte V, Zhai P, Pörtner HO, Roberts D et al (2018) An IPCC special report on the impacts of global warming of 1.5°C above pre-industrial levels and related global greenhouse gas emission pathways, in the context of strengthening the global response to the threat of climate change, sustainable development, and efforts to eradicate poverty. Intergovernmental Panel on Climate Change. https://www.ipcc.ch/sr15
27. Krey V, Masera O, Blanford G, Bruckner T et al (2014) Climate change 2014: mitigation of climate change. Contribution of working group III to the fifth assessment report of the intergovernmental panel on climate change annex II: metrics and methodology. Cambridge University Press, New York

28. Denholm P, O'Connell M, Brinkman G, Jorgenson J (2015) Overgeneration from solar energy in California: a field guide to the duck chart. National Renewable Energy Laboratory. NREL/TP-6A20-65023. https://www.nrel.gov/docs/fy16osti/65023.pdf
29. Breuer H (2018) A microgrid grows in Brooklyn. Siemens AG [Online]. https://www.siemens.com/innovation/en/home/pictures-of-the-future/energy-and-efficiency/smart-grids-and-energy-storage-microgrid-in-brooklyn.html. Accessed 19 Jan 2019
30. Gajda J (2017) The North Carolina solar experience: high penetration of utility-scale DER on the distribution system. Presented to IEEE power & energy society working group on distributed resources integration, Jul 2017
31. Anair D, Mahmassani A (2012 June) State of charge: electric vehicles' global warming emissions and fuel-cost savings across the United States. Union of Concerned Scientists. https://www.ucsusa.org/sites/default/files/legacy/assets/documents/clean_vehicles/electric-car-global-warming-emissions-report.pdf
32. Krauss C (2016) Coal production plummets to lowest level in 35 years. New York Times
33. Kuykendall T, Cotting A (2018) Coal plant closings double in Trump's 2nd year despite 'end of war on coal'. S&P Global Market Intelligence
34. Bruninga R (2012) Overlooking L1 charging at-work in the rush for public charging speed. United States Naval Academy, Annapolis
35. Siemens AG (2016) How microgrids can achieve maximum return on investment: the role of the advanced microgrid controller. Siemens AG [Online]. https://w3.usa.siemens.com/smartgrid/us/en/microgrid/Documents/MGK%20Guide%20to%20How%20Microgrids%20Achieve%20ROI%20v5.pdf. Accessed 19 Jan 2019

Part III
Science, Technology, and Innovation

The Evolving 5G Landscape

Liam Quinn

Contents

1 Introduction: Overview of 5G

The "G" in wireless telephony is nominally associated with the generational changes in technologies and services in the carrier domain. The current iteration of 4G is a well-understood evolutionary implementation of the wireless mobile telecommunications network, enabling the wireless Internet and cloud capabilities available on contemporary mobile handsets and devices. Extending this framework, the fifth generation of the mobile broadband network, or 5G, will drive four primary usage models:

1. Enhanced Mobile Broadband (eMBB)
2. Ultra-Reliable Low Latency Communications (uRLCC)
3. Massive Machine Type Communications (mMTC)
4. Ultra-High Speed, Low Latency Communications (uHSLLC)

L. Quinn (✉)
Dell Technologies Inc., Austin, TX, USA
e-mail: liam.quinn@dell.com

© Springer Nature Switzerland AG 2020
S. McClellan (ed.), *Smart Cities in Application*,
https://doi.org/10.1007/978-3-030-19396-6_7

To achieve these functional models, the standards body for cellular communications (3GPP) has specified additional spectrum for 5G to support diverse workloads, as shown in Fig. 1. The figure provides a general overview of 5G "Phase 1" and "Phase 2" application characteristics in terms of data rate and latency. Phase 1 defines several new usage scenarios including eMBB, uRLLC, and mMTC which enable, improve, or extend contemporary applications. Next-generation applications enabled by new service classes, such as Ultra-High Speed Low Latency Communications (uHSLLC), could emerge in Phase 2 and beyond. As a result, 5G will provide a "hyperconnected environment" where data will be accessible to users in a reliable, flexible, and secure way independent of the device type, underlying architecture and OS environment [4, 5]. Software-Defined Networking (SDN), Network Function Virtualization (NFV), and virtualization of compute resources will drive changes to fundamental principles of the service provider's network structure, enabling a more decentralized architecture.

Applications which leverage the improvements to mobile broadband of eMBB include Augmented/Virtual Reality (AR/VR), cloud-based remote office functionality, and high-definition or broadband audio/video, as well as certain mission-critical and complex remote applications.

Applications which leverage the enhanced support for machine-to-machine (M2M) communications of mMTC include various approaches to the Internet of Things (IoT) as well as "Smart Home" and "Smart City" technologies which range from home security to intelligent water management to automated pollution

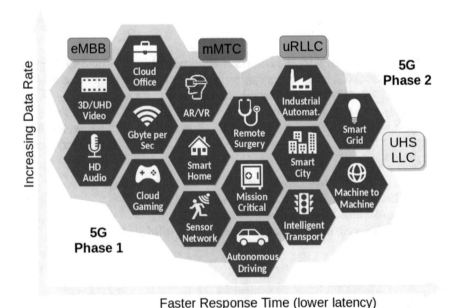

Fig. 1 Primary usage models of 5G, with examples of applications. Adapted from [1–3]

management and other applications important to the management of growing municipalities.

Applications which leverage the reliability and low latency of uRLCC include autonomous vehicles and intelligent transportation, including "V2X" or vehicle-to-everything communications, real-time, remote healthcare applications ("telemedicine"), and the emergence of the tactile Internet.

The flexibility of the virtualized architecture of 5G will also disrupt the enterprise datacenter, providing flexibility in network configuration and load balancing based on usage models, traffic types, and end-point mobility. Radical new use cases for 5G-capable systems will impact future designs for client devices, in areas such as power/spectrum management, optimized user experience across heterogeneous networks, the "always on" potential of cloud-based storage, and the "always-changing" potential of fog-based wireless interaction. As a result, new requirements for client devices will emerge to meet the rapidly changing needs of consumers and businesses [6], as shown in Fig. 2.

While 5G is not yet fully specified, there are several areas of intense activity, including end-to-end latency requirements, in-building service requirements, spectral reuse and propagation issues, and the always-present "Quality of Service" (QoS) problem. While standards are being finalized, the industry will see a phased, multi-year deployment of 5G with end devices (smart handsets) launched in trials with major carriers [7]. The following sections discuss some of the most important aspects of 5G, including issues associated with applications/services, air interface/spectrum requirements, and virtualization/software-defined subsystems.

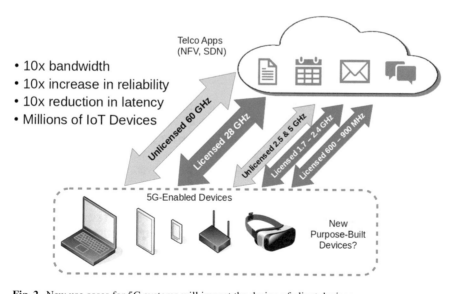

Fig. 2 New use cases for 5G systems will impact the design of client devices

1.1 Mission-Critical Services

Connecting mission-critical industries with ultra-reliable communications is a key driver of 5G. 5G will connect new industries and enable new services including mobile health and telemedicine systems, public safety/disaster alert, remote control of heavy machinery, factory automation, sustainable cities, and "smart" electrical generation and distribution.

Improved network capabilities will drive new use cases around mission-critical services and enterprise requirements, and will incorporate highly mobile applications. The planned high bandwidth, low latency features of 5G will drive more compute to the edge in a "distributed virtual cloud," enabling richer end user experiences, data analytics, content delivery, and value added services in addition to mission-critical connectivity for new or enhanced municipal services.

1.2 Software-Defined Infrastructure

Network function virtualization (NFV) and software-defined networking (SDN) tools and architectures are enabling service providers to reduce network costs, simplify deployment of new services, and scale network growth and expansion. Other initiatives include the virtualization of the radio-access network, as well as development of cloud radio-access network architectures. This enabler addresses the need for computing beyond the edge and from the cloud infrastructure and the need for software solutions to enable such a service. Enhanced security and data management architectures will also play an important role for this enabler.

The software-defined infrastructure "stack" is typically composed of several interoperating hardware and software layers. These layers include the *physical infrastructure*, one or more *virtualization layers*, a suite of *software-defined capabilities* or functions, and an overarching *management function* which binds the layers of the stack together [8].

The *physical layer* of the stack leverages a variety of "real" devices and systems, including computers (servers), network storage systems, network-resident nodes (switches, routers), as well as the operating systems, firmware, subsystems, and other software components necessary for proper function in a networked environment, and for application-hosting purposes. The software-defined infrastructure which leverages these underlying systems is not directly dependent on specific instances of the hardware, so the supporting infrastructure can be managed, scaled, and deployed independently of the overarching, software-defined system.

The *virtualization layer* of the stack leverages a variety of software components which overlay the physical infrastructure systems and their associated subsystems. This abstraction presents the heterogeneous collection of resources as a homogeneous or "virtual" distributed computing system, complete with all necessary elements required for full operation. Virtualization software is loosely divided into two functional paradigms: complete system virtualization, and application-level virtualization facilities. Complete system virtualization generally provides a selection of well-known, emulated hardware/firmware capabilities, upon which a user or administrator can "install" unmodified operating system(s) and application image(s) which effectively realize a fully functional computer system. Examples of complete system virtualization include the Linux Kernel Virtual Machine (KVM) [9] and the VMWare software suite (e.g., vSphere and the VMWare NSX suite) [10]. Application-level virtualization generally provides a selection of well-known, emulated operating system capabilities, upon which a user or administrator can "run" specially packaged application(s) as long as the application package (or "container") includes all necessary software features (libraries, etc.) which enable the complete function of the application. An example of application-level virtualization facility is Docker [11].

The *software-defined capabilities* of the stack leverage a collection of emulated compute, storage, and networking systems which are implemented by the virtualized layer of computing resources. Automated and semi-automated telemetry, control, and configuration algorithms are deployed as a critical part of the software-defined capabilities. This "orchestration" of the emulated devices, via the virtualization layer, is capable of automatically transforming the underlying resources to respond to contextual requirements, or to enable new/different applications. System context may be prescribed by users via direct configuration of emulated systems, or by abstract description of performance or operating parameters. Deployment, configuration, and monitoring of the infrastructure and virtualized resources is then effected by the orchestrator in much the same fashion as a conductor manages a symphony orchestra in playing sheet music.

The *management function* of the software-defined infrastructure isn't a "layer." Rather, it's a vertical feature which interfaces with all layers, and is common to all telecommunication or computer networks. Management functions typically include a collection of vendor-specific user-interfaces which are used to define operational parameters, deploy and configure software components, and enable provisioning of specialized resources. Management functions also typically include a "marshaling" capability which enables a form of single-point control. In this fashion, management function ensures that necessary infrastructure operations are completed according to desired standards, and that service-level-agreements (SLAs) and performance are maintained during system operation. In some software-defined architectures, embedded configuration facilities for physical and/or virtual systems are abstracted so that a common management interface is presented to the user [12].

1.3   5G Air Interface

The 5G network will support a heterogeneous set of air interfaces, from evolutionary derivatives of current technologies (such as Wi-Fi and Mobile broadband) to new technologies and network architectures which haven't been developed yet.

Challenges remain on the coexistence of Wi-Fi (unlicensed spectrum) with 5G (licensed spectrum) and the potential for deployment of 5G into unlicensed bands (think of this as 5G NR-U). This new connectivity standard will provide multi-gigabit data throughput by using carrier aggregation, massive multi-input/multi-output antennas (MIMO), and simultaneous use of licensed and unlicensed spectrum. Seamless handover between heterogeneous wireless access networks will need to be a differentiated feature of 5G, as well as use of simultaneous radio access technologies to increase reliability, connectivity, and availability.

As part of the overall development of 5G, the 3GPP[1] defined a 5G NR (new radio) specification which operates in spectrum below 6GHz (sub-6GHz ... which includes 4G technologies), and above 24 GHz, in the so-called mmWave bands. Operation in the mmWave bands will require the deployment of large numbers of new, small-footprint base stations which are capable of delivering high bandwidth payloads with extremely low latency (<1 ms)[2] to end-point devices and applications. While mmWave technologies can deliver very high data rates, optimizing radio-frequency propagation in these bands is a complex challenge.

Clearly, substantial work remains in the development of standards and coexistence schemes to drive a seamless "intelligent connected end device." This paradigm will be central in the migration from Wi-Fi and 4G to 5G and a truly pervasive mobile connected environment.

1.3.1 5G as a Catalyst

The requirement for 5G wireless connectivity will provide a catalyst for the convergence of key technology transitions such as Artificial Intelligence (AI), Machine Learning (ML), Augmented Reality (AR), Virtual Reality (VR), the Internet of Things (IoT), and edge-based computing. To enable these applications, the NR specification will be band agnostic—meaning it can be deployed on low, medium, and high bands with no restrictions. The first wave of networks and devices will be classed as Non-Standalone (NSA). The 5G Standalone (SA) network and device standard is still under review by 3GPP, and will emerge in future revisions of the standard, which are known as "Releases." 5G Releases are briefly described below, and the network transition is diagrammed in Fig. 3.

Release 15 (June, 2018): In R15, additional new spectrum, new protocols, client devices will connect to (new) 5G frequencies for data throughput leveraging the existing 4G packet-core infrastructure. 5G (NSA and SA) will provide increased data bandwidth and connection reliability via two new spectral allocations. As discussed previously, R15 overlaps and extends 4G LTE frequencies from 450 MHz to

[1] 3GPP, or the "Third Generation Partnership Program," is the federated, system-level industry standards organization for wireless telecommunications. More information at https://www.3gpp.org.

[2] For comparison, today the QoS Class Identifiers used in LTE specify packet delay values between 50 ms and 300 ms.

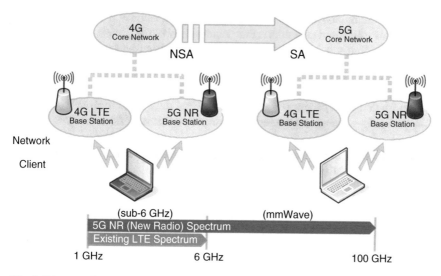

Fig. 3 Diagram of 5G network transition, spectrum, and connections

6 GHz in the sub-6GHz bands, and extends from 24.25 GHz to 52.6 GHz in the mmWave bands.

Release 16 (2019–2020 timeframe): In R16, The 5G network will extend to cover the low-mid and high bands (sub-6GHz and mmWave spectrum) and will be backward compatible with the NSA Network and device deployments. With this Release, carriers will migrate from 5G NSA to an all 5G SA infrastructure in a fashion transparent to the end-users, as diagrammed in Fig. 3.

1.3.2 Coexistence of Licensed and Unlicensed Networks

The expansion of wireless broadband access network deployments coupled with heavy video service usage is resulting in an increased scarcity of available radio spectrum. Cellular technologies and wireless local area networks will need to exist in the same unlicensed spectrum bands. The two most prominent technologies (LTE and Wi-Fi) are based on standards designed to operate in different part of the frequency spectrum and were not designed to coexist in the shared band. A previously fragmented licensed wireless industry has consolidated globally on LTE as the defacto standard, and wireless Ethernet (Wi-Fi) has dominated unlicensed deployments due to cost and simplicity. 5G will be designed to integrate with LTE networks, and many 5G features may be implemented as LTE-Advanced extensions prior to full 5G availability.

Furthermore, "carrier aggregation" (a key LTE-Advanced feature) optimizes available spectrum more effectively, increasing capacity and increasing throughput. Deployment of LTE in unlicensed frequency bands will enable "neutral hosts" to offer "Enterprise LTE" as a service to enterprises and large campuses such as public venues, hospitals, and industrial facilities. A primary concern in coexistence of licensed and unlicensed networks is that devices capable of "license-assisted access" (LAA) will unfairly use the 5GHz band and other unlicensed bands due to the following reasons:

- Not implementing "listen before talk" (LBT) algorithms. For example, LTE-U does implement a carrier sense algorithm to set duty cycle for transmission but after then implements a fix interval transmission that can effectively jam a Wi-Fi signal thereafter.
- Parameters for the interval between carrier sensing and transmitting and duty cycle can be set aggressively, causing LTE to gain unfair access to shared spectrum.
- Listen level (i.e., carrier sense thresholds) for LTE-U implementation is set in the -62 dm range, while Wi-Fi access points listen down to -85 dBm. Thus it can be said Wi-Fi access points are more considerate of other, lower powered Wi-Fi access points.

1.3.3 Spectrum Reuse

The continued increase in the number of smart mobile devices and the emerging classes of IoT smart end points is driving the need for additional spectrum for cellular communication. Cellular spectrum allocation to each carrier is fixed, which requires new approaches to spectrum sharing to meet demand. As a result, 3GPP is supporting research in three key areas: (1) Spectral efficiency improvement, (2) Higher network cell density, and (3) Evaluation of underutilized spectrum resources.

These research areas require radio resources and mobility management techniques across both homogeneous and heterogeneous spectrum which increases both system complexity and cost. Spectrum sharing may occur between an incumbent commercial usage and a secondary commercial usage. An example of this is TV White Space (TVWS) as a commercial use, sharing this with Wi-Fi usage in a public–private application. In the case of 5G networks, governments, wireless vendors, carriers, operators, and standards associations worldwide are driving toward a common solution. However, with the range and scope of the use cases and functional requirements, network operators will most likely need to deploy a set of solutions across selected low and high band spectral allocations, which would include exclusive-use licensed, shared-access, and unlicensed spectrum.

All radio spectrum is reused in some way and historically, regulatory authorities have managed spectrum by organizing it into bands (fixed, land, mobile) with specific usage, and then channels within bands. Essentially there are a number of methods for spectrum reuse:

- Frequency as the primary reuse mechanism parameter with output power limitations, and transmission masks as additional reuse parameters. Access protocols provide real-time reuse using frequency division multiple access (FDMA).

- Spatial reuse leverages geographic separation as the primary parameter using controlled radiation patterns.
- Temporal reuse uses time as the primary parameter, using time slots with access protocols for time division multiple access (TDMA).
- Code space reuse uses orthogonal pseudorandom spreading sequences for code division multiple access (CDMA).
- Hybrid reuse—which combines space, time, and frequency parameters.

5G will likely use new hybrid combinations of these reuse techniques, combining cognitive radio technologies and geolocation-based techniques for spectrum sharing. Additionally, 5G networks will require additional spectrum in both low and high bands, more small cells and more ad-hoc small cells with multiple technologies. As a result, LTE Advanced will continue to evolve in bands below 6 GHz, in a backwards compatible way. In parallel, new radio access technologies will emerge in cm/mmWave bands (e.g., 28 GHz, 37–40 GHz, and 60+ GHz) and in mid-band microwave (3.4–4.2 GHz) with no backward compatibility constraints [13].

2 Attributes, Use Cases, and Market Drivers

In general, 5G networks are being designed to handle more connections, faster speeds, and lower latencies than conventional network architectures. These key attributes, as well as other important characteristics, are summarized in Fig. 4.

As a simple qualitative comparison with current-generation wireless technologies, the 5G network will handle 100× more endpoints (over 20 billion user devices and over 1 trillion terminals), and will offer a 100× increase in peak theoretical speeds (more than 10 Gbps[3]). These improvements in service will necessarily have to be delivered in a highly reliable fashion ("five 9's," or less than a few seconds of downtime per year), and at high ground speeds (velocities >300 miles/h). This is quite a tall order!

Additional key requirements addressed by 5G technologies include:

- 10× reduced end-to-end latency (<5 ms end-to-end latency, <1 ms over-the-air latency),
- 20× higher user data rates over 4G (5—20 Gbps peak data rates),
- 1000× higher mobile data volume per area,
- uniform experience regardless of user location ("edgeless"), and
- 10× longer battery life for low power machine-to-machine communications.

For a specific technology reference, Table 1 outlines the relative characteristics for each of the three primary communication media types over the past couple of decades. Note that 5G is comparable to 10 Gbps wired Ethernet which is

---

[3] Gbps = gigabits per second.

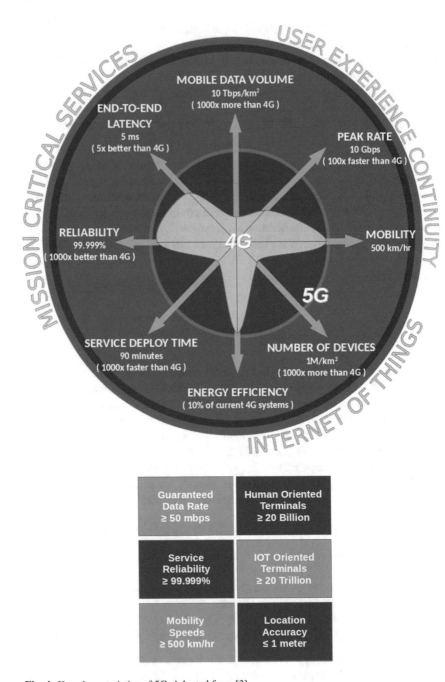

Fig. 4 Key characteristics of 5G. Adapted from [3]

Table 1 Reference points for networking technologies

Media types	Ethernet		Wi-Fi		Cellular	
	Standard	Data rates	Standard	Data rates	Standard	Data rate
1980s	802.3	10 Mbps			1G	24 Kbps
	802.3u	100 Mbps				
1990s	802.3ab	1 Gbps			2G	500 Kbps
			802.11b	11 Mbps		
2000s	802.3an	10 Gbps	802.11 g	54 Mbps	3G	2 Mbps
			802.11n	600 Mbps		
2010s	802.3bj	100 Gbps	802.11 ac	850 Mbps	4G	100 Mbps
			802.11ad	3 Gbps		
2018/20s	802.3bs	400 Gbps	802.11ax	1.2 Gbps	5G	10 Gbps
			802.11ay	7 Gbps		

currently used in today's datacenters, and at 10 Gbps a 1.25 GB movie downloads in roughly 1 s.

5G is positioned to support a fully mobile and connected environment. Due to the "always-on" context, socio-economic transformations will be enabled across all segments of society. Technologies from both the traditional carrier domain (licensed spectrum) and non-carrier domains (unlicensed spectrum[4]) will be united in a single, end-to-end networking paradigm. As a result, the business model for 5G is extremely complex. Conventional network operators will not be able to charge customers for increasing data usage via current air interfaces and bandwidth-limited networking infrastructures. The emerging model for monetization of 5G is likely to include partnerships and content delivery services with extension to other industries through the expansion of the IoT vertical markets (e.g., automotive, transportation, smart city infrastructures, etc.).

Key opportunities in this new mobile broadband market will include leveraging the multitude of devices available in this "network of networks" to transport data and content in the most cost-effective and efficient manner, based on applications, payload, and latency requirements. The diversity of some of the use cases driving the 5G environment is shown in Fig. 1. Additionally, the commoditization of many important networking technologies as well as the move toward virtualization of compute and network functions enables non-traditional equipment suppliers to play a larger role in the telecommunications ecosystem. As a result, 5G network rollouts will feature an increasingly larger role for original equipment manufacturers (OEMs) from the Information Technology sector (IT). This "hybrid" architecture and ecosystem is diagrammed in Fig. 5.

Additionally, annual surveys of telecom operators [14] reveal that the general areas of mobile connectivity, IoT and machine-to-machine (M2M) connectivity, and low latency applications are the leading drivers for 5G use cases. Not surprisingly,

---

[4]Wi-Fi, WiGig, and other emerging technologies.

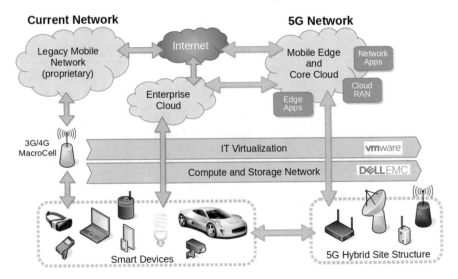

Fig. 5 5G will enable IT OEM suppliers to provide a larger portion of the infrastructure

security concerns vary widely for common use cases, with heightened security demands and challenging implementation requirements expected for ultra-reliable, low latency, and M2M scenarios. In general, network service providers agree that enhancements to mobile broadband, massive IoT/M2M requirements, and applications requiring ultra-reliable/ultra-low latency services will be significantly transformed by the capabilities of 5G networks.

Based on these technical capabilities and features, the emerging use cases that can take advantage of features can be broken down into three key areas that are central to the vision for 5G, and which are discussed in more detail in the subsequent sections. Additionally, a fourth key area will arise as the volume of information necessarily increases—artificial intelligence, machine learning, and data analytics.

1. Mobile Connectivity
2. Internet of Things (IoT)
3. Mission-Critical Services

2.1 Mobile Connectivity

Mobile connectivity continues to be one of the main drivers for the next-generation broadband network. As the demand for mobile connectivity increases at an astounding 50–60% compound annual growth rate (CAGR), the mobile broadband requirements for 5G extend far beyond basic Internet access and covers connectivity in very

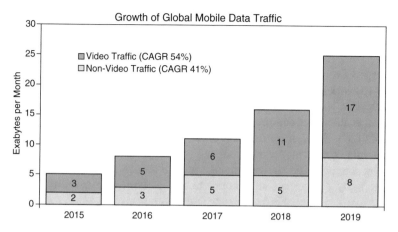

Fig. 6 Predicted mobile data traffic growth toward 2020, in exabytes per month. The trend shows a combined 53% compound annual growth rate (CAGR). Adapted from [15]

dense environments, ultra-high speed connectivity, and access to rich media anytime, anywhere. According to most sources, and as depicted in Fig. 6, the growth in mobile data traffic will be driven by the popularity of applications with requirements for video streaming [16], including ultra-high-definition media, virtual conferencing, immersive virtual reality, tactile Internet, and immersive gaming.

Reliable and persistent connectivity in challenging situations such as high mobility applications, very dense or sparsely populated areas, and the journeys covered and supported by heterogeneous networks and technologies. As content continues to migrate to the Internet, 5G will accelerate the mobile Internet for delivery of multimedia, voice, video, and services in this new hyperconnected wireless environment [5]. 5G addresses these challenges through the paradigm of "eMBB," or enhanced Mobile Broadband which leverages LTE installed infrastructures (voice/control).

With the increased bandwidth and low latency features of a 5G mobile network environment, service providers will take advantage of these features to deliver better interactive experiences with real-time applications that will run on the edge of their networks. It will be possible to leverage software-defined Wide Area Networks (SD-WANs) and network function virtualization software to deliver services from virtual data centers hosted in the cloud. The high bandwidth features and low latency of 5G will support enhanced features and applications such as AR/VR, IoT, and real-time multimedia and location based services that are deployed at the edge of a service provider's network.

It is important to remember that devices connected to 5G networks are not going to be less intelligent than they are today (perception is that all things will come from the cloud given data rates/low latency), with less storage and computing capabilities. With the addition of more AI/ML features to end-point/client compute devices, 5G capable devices will have more advanced features and computing capabilities than they have today, with more sensor and intelligence that enables them to become more powerful and capable edge devices. More intelligence, more computing will migrate from the cloud down to the edge devices, driving more demand for local storage (big data analytics), more sensors, and intelligence on these devices [15].

Converged Compute and Mobility

As 5G networks deliver very high throughput and very low latency communications, devices will be able to leverage distributed and cloud computing solutions. This will enable more compact and thin end user devices.

In addition, deployment of small cells in dense campus environments together with mobile edge compute solutions will enable local LTE networks used exclusively by enterprises and large campuses (public venues, hospitals, industrial facilities, etc.). By converging local compute and carrier-grade mobility "Enterprise LTE" networks will enable low latency and high privacy networking solutions.

Market Landscape:

- Enterprise mobility market is set to rise at 25% CAGR through 2022 as more enterprise users prefer device flexibility.
- Enterprises will have to deliver 300% more wireless access points to provide future Internet performance that is similar to the performance in the pre-BYOD era.

Enhanced Video

5G Networks are expected to deliver 10× more bandwidth which will drive enhanced video experience in various form factors and enable 360 Conferencing, UHD Streaming, and VR applications.

Market Landscape:

- By 2019, ~70% of global Internet consumption will be video content.
- 4 K technology market is expected to grow at a CAGR of ~22% from 2015 to 2020.
- Cloud-based video conferencing market to grow at a CAGR of ~40% by 2019.

Connectivity Everywhere

5G is expected to connect low ARPU areas through non-traditional access points, such as drones, balloons, satellites, operating in both unlicensed and licensed spectrum. 5G networks will also be able to increase the network capacity in dense areas, like stadiums, airports, open air assemblies, etc., by enabling more spectrum for access, deploying small cells, and dynamically managing the network resources via leveraging virtualization and SDN technologies together with mobile edge computing.

Market Landscape:

- There are still >4 billion people not connected to the Internet, most of them in the developing world.
- By 2020 mobile Internet penetration rates in developing markets will reach 45% from 28% today. However, most of the growth will be in small form factor handsets.
- By 2020, small cells are expected to carry a majority of traffic with overall data volume expected to grow up to 1000 times compared to 2010.

Seamless Mobility
Mobile devices moving between heterogeneous networks while maintaining continuous connection will be a key requirement for 5G networks. Heterogeneous network technologies with considerable emphasis include disparate unlicensed networks (e.g., Wi-Fi) and multiple tiers of licensed cellular networks (Macro, Mico, and Femto cells). These network paradigms all utilize various types of physical-layer communications schemes, media access controls, and management architectures.

To-date, mobility in Wi-Fi networks has been limited, although newer technologies are being developed to address issues such as roaming (802.11r), FILS (802.11i), and seamless authentication (Passpoint). Multi-layer connection control based on end device speed will allow all endpoints to become more efficiently connected to the broadband network, and minimize the need for handoffs of a single connection between multiple network instances.

Market Landscape:

- Smart connection managers that can aggregate traffic across licensed and unlicensed networks as well as select optimal connection contexts will help system and device providers differentiate via improved connection experiences.
- These new smart connection managers will also improve connection resilience and reliability for IoT devices.

2.2   Internet of Things

The ability of 5G networks to connect massive number of endpoints is targeted for the growing Internet of Things (IoT) market, which is dominated by "headless" devices tasked with automating various processes. 5G provides a robust platform to support a massive number of sensors and actuators with stringent energy and transmission constraints. The delivery of IoT systems, characterized by the need to integrate the management of massive number of connected devices, will continue to drive the growth and expansion of the 5G network, offering a potential economic impact of $4–11 T a year in 2025 [17], with most applications dependent on cellular connectivity.

Some of the key use cases driving 5G technology in IoT scenarios include connected cars and autonomous vehicles, smart homes, sensor networks, surveillance applications, continuous health monitoring, infrastructure monitoring, and modernization of utilities and municipal services. These challenges are addressed through the 5G paradigm of "mMTC," or massive Machine Type Communications.

Intelligent Transportation Systems
Highly reliable and low cost 5G technologies will enable connecting transportation systems for logistics and vehicle-to-vehicle communication applications.

Market Landscape:

- A quarter-billion connected vehicles will be on the road by 2020, with new vehicles increasing the proportions of connected cars.

- Embedded connectivity and multiple types of network options will dominate the connected car market, making automobiles effectively mobile data and entertainment centers.

Smart Society

5G networks will support long range, low cost, and low power technologies that will enable smart city applications, such as intelligent traffic control and smart parking solutions.

Market Landscape:

- Regulatory initiatives such as smart meters estimate an 80% penetration by 2020 are driving rapid adoption of data analytics for municipal applications. The increasing number of intelligent vehicles will propel initiatives of this nature.
- Services in the municipal space result in low average revenues per-user (ARPU) and are generally hard to scale without appropriate partnerships. As a result, additional public/private partnerships for virtualized infrastructure services may emerge.

Sensor Networks

Low power and low cost solutions driven by 5G networks will enable cellular connectivity on devices, such as wearables, smart meters, and environment monitoring sensors.

Market Landscape:

- Shipments of smart wearables will continue their rate of growth (28% CAGR since 2015), exceeding the estimated 215 M units in 2019.
- This is a highly crowded market with several players from Google (smart glass, Nest thermostat) to location service providers. The data produced by sensors and other devices connected to the 5G network will propel new approaches to data mining and analytics.

2.3 Mission-Critical Services

Some important applications and usage models require very high levels of reliability, global coverage, and/or very low latency. These areas have traditionally been supported by dedicated, application-specific networks, such as public safety and emergency services. 5G will provide a more efficient platform for connecting existing mission-critical services such as mobile health and telemedicine systems, public safety/disaster alert applications, remote control of machinery or systems in dangerous conditions, factory automation, municipal sustainability initiatives, and the Smart Grid. The 5G network's use of mmWave spectrum will reduce the latency and improve response time of new mission-critical operations such as autonomous vehicle management, remote robotic surgery, and precise location services. These challenges are addressed through the 5G paradigm of "uRLLC," or Ultra-Reliable Low Latency Communications.

High Reliability Communications

5G networks will deliver 10× reduction in end-to-end latency, peak data rates of 10 Mbps, higher reliability, and support for mobility applications up to 500 km/h. This will enable more robust communication services for disaster recovery and public safety applications.

Market Landscape:

- The market for disaster recovery as a service (DRaaS) is estimated to grow from $1.4 Billion in 2015 to over $12 Billion in 2020, at a CAGR of 53%. This market will emerge as a niche segment, with high margins.
- Rugged device form factors are most suitable for this usage model, with specialized software, registration, and interconnection services increasing due to management/monitoring requirements.

Industrial Controls

Flexible network slicing and virtualization of the network core will enable 5G networks to support diverse workloads and applications based on the needs of bandwidth, latency, and capacity. The Telco and service provider networks will need to transform to an open virtualized flexible and agile models for these and new workload types.

Tele-Health

High data rates, low latency, and high reliability networks enabled by 5G networks will enable new applications, such as tele-health applications, such as remote surgery, distant care, and support for a tactile Internet.

Market Landscape:

- The market for tele-health applications will exceed $35B by 2020, growing at a CAGR of over 14% between 2015 and 2020. The mobile health market will increase beyond $18 B, growing at an estimated 40% for the next several years.
- Regulations and highly customized needs make tele-health a specialist segment, with high margins and substantial governmental oversight providing additional data analytics opportunities.

2.4   *AI/ML and Big Data Analytics*

There are a number of key emerging technologies that are changing the information and communications ecosystem—including 5G, Wi-Fi6, AI/ML, AR/VR, and IoT. 5G is a pivotal platform that will drive and enable a hyperconnected society and accelerate innovation and widespread adoption of these technologies and applications and the combination of these over the next number of years will have a dramatic influence on innovation and vertical industries that depend on the IT and telecom industries and services—such as user experiences in communication, human-to-machine interaction, real-time applications, multimedia, and digital content generation and delivery.

Market Landscape:

- The scale required for IoT will drive a longer term requirement for additional capacity as well as bandwidth on demand. Quality of Service (QoS) requirements for many IoT-enabled services will produce new avenues for value and differentiation.
- The transition to a software-defined infrastructure will leverage the capabilities of compute, networking, and storage virtualization to drive new capital and operational business models. Simultaneously, infrastructure changes based on 5G mmWave features will lead to a transformation of municipal network and carrier business models.

3 Conclusions

5G will be the first end-to-end cellular networking architecture which is fully software-defined from the edge radio though to the core. It has being designed to provide significant increases in mobile bandwidth, low latency, network capacity enabled by a disaggregated, virtualized network architecture [18]. These critical architecture features form the network platform for new innovative business, operational, and technology outcomes across all segments of society enabling a more connected environment between humans, machines, and things.

References

1. Teyeb O et al (2017) Evolving LTE to fit the 5G future. Ericsson Technology Review
2. Sakaguchi K et al (2017) 5G-MiEdge: millimeter-wave edge cloud as an enabler for 5G ecosystem. Deliverable Horizon2020, EU Contract No. EUJ-01-2016-723171, 5G-MiEdge D1.1. https://5g-miedge.eu/
3. 5GPP (2015) 5G vision: the 5G infrastructure public private partnership: the next generation of communication networks and services. https://5g-ppp.eu/wp-content/uploads/2015/02/5G-Vision-Brochure-v1.pdf
4. RCR Wireless News (2019) The journey to 5G: communications and computing will converge in the clouds. https://www.rcrwireless.com/20190219/wireless/the-journey-to-5g-converging-communications-and-computing
5. Dell Technologies Digital Transformations (2019) 5G creates new opportunities – what will be the Uber of tomorrow? https://www.delltechnologies.com/en-us/perspectives/5g-creates-new-opportunities-what-will-be-the-uber-of-tomorrow/
6. Quinn L (2019) 5G network landscape: the evolving 5G mobility landscape and its impact to the medical industry (unpublished whitepaper)
7. Flore D, Bertenyi B (2015) Tentative 3GPP timeline for 5G. http://www.3gpp.org/NEWS-EVENTS/3GPP-NEWS/1674-TIMELINE_5G
8. Raza M (2018) What is software-defined infrastructure? SDI explained. BMC Blogs. https://www.bmc.com/blogs/software-defined-infrastructure/
9. Linux Kernel virtual machine. https://www.linux-kvm.org/page/Main_Page
10. VMWare. https://www.vmware.com/

11. Docker. https://www.docker.com/
12. Wickboldt J et al (2015) Software-defined networking: management requirements and challenges. IEEE Comm Mag 53:278–285. https://doi.org/10.1109/MCOM.2015.7010546
13. Press Release (2016) FCC adopts rules to facilitate mobile broadband and next generation wireless technologies in spectrum above 24 GHz. https://apps.fcc.gov/edocs_public/attachmatch/DOC-340301A1.pdf
14. Marshall P (2017) 5G operator survey: a TIA market report. TIA Online. https://www.tiaonline.org/wp-content/uploads/2018/05/5G_Operator_Survey.pdf
15. Bangerter B et al (2014) Networks and devices for the 5G era. IEEE Comm Mag 52(2):90–96. https://doi.org/10.1109/MCOM.2014.6736748
16. Barnett T et al (2015) Cisco visual networking index (VNI) mobile data traffic update, 2015–2020 Cisco Knowledge Network (CKN). Cisco
17. Manyika J et al (2015) The internet of things: mapping the value beyond the hype. McKinsey Global Institute. https://www.mckinsey.com/~/media/McKinsey/Business%20Functions/McKinsey%20Digital/Our%20Insights/The%20Internet%20of%20Things%20The%20value%20of%20digitizing%20the%20physical%20world/The-Internet-of-things-Mapping-the-value-beyond-the-hype.ashx
18. Int'l Telecommunications Union (2015) IMT Vision: framework and overall objectives of the future development of IMT for 2020 and beyond. ITU-R M.2083-0. https://www.itu.int/rec/R-REC-M.2083-0-201509-I/en

Architecting IOT for Smart Cities

Achamkulamgara Arun

Contents

1 Introduction

As the physical world continues to fuse with the digital world, cities populated with communities of people, businesses, and government entities cannot dismiss the transformational possibilities to drive new value. On a daily basis cities are challenged to develop economic growth, improve public safety, enhance public transportation and infrastructure while reducing financial costs and their environmental footprint. These cities need to tap into innovative ways to leverage the power of technology and value of data to address these challenges and create a sustainable pace for change that can enable a better future.

A. Arun (✉)
IOT Product Solutions, Cognizant, Austin, TX, USA
e-mail: Achamkulamgara.Arun@cognizant.com

© Springer Nature Switzerland AG 2020
S. McClellan (ed.), *Smart Cities in Application*,
https://doi.org/10.1007/978-3-030-19396-6_8

The Internet of Things (IoT), sensors, big data, cloud platforms, and services are helping to transform every facet of the world, including cities, and are at the center of a digital movement. While these new technological advances seek to enrich people's lives with connecting and optimizing productivity, delivering new services and support, nations are concerned more in enriching the lives of their people and solving day to day, common issues. Connected places like smart homes, offices, and buildings or stadiums are getting tooled as part of the larger ecosystem initiative, namely smart city. Smart city programs have focused on improving the efficiency and experience with deploying smart solutions for public transportation, energy, water, trash, parking, and public safety. To adapt a journey towards adopting IoT, every digitalization program smart city must overcome challenges emerging from complexity of sensors, connectivity and enterprise IoT, and cloud technologies. Recent studies [1] indicate top concerns for smart city councils are around security, ethics, and resiliency. The pervasiveness of surveillance like facial recognition and image, data analytics has pushed privacy, ethics, and AI issues to the forefront. Several governments, regional, and city wide variation of policies across geographies are expected in the next several years to tackle these trends.

In this chapter, we look at various architectural and functional capabilities that are required for a smart city initiative. An overview of the recommended architecture and implementation technology choices, typical constraints in implementations, terminology, and technology principles are highlighted. We attempt to provide various levers around non-functional needs like city or state governance, meeting compliance and regulatory considerations. These are applicable at various intersections, whether they develop custom solutions or standardized solutions using open, federated platforms and/or commercial platforms.

A reference architecture provides a means from choosing the right sensing element, connectivity (network, communication, interconnected, and so on), and integrating with IoT and cloud platforms. Based on years of IoT experiences, architecture development and implementation, system integrators, product companies, and enterprises along with multiple ecosystem players are addressing the needs of open and scalable approach to realizing smart city enterprise IoT solutions through repeatable, reliable, architecture patterns. We take you through many of the smart city architecture considerations.

Figure 1 presents a detailed Smart City ecosystem covering various aspects of technology and implementation. However, the Smart City enterprise architecture can be layered across the following general areas:

- City Architecture
- Enterprise IoT Architecture
- Operating Architecture

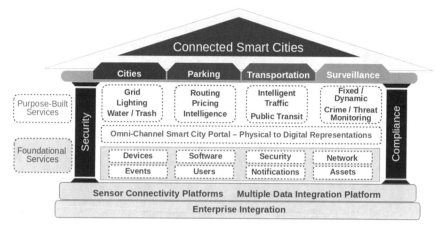

Fig. 1 Detailed smart city ecosystem

1.1 City Architecture

The City, region, and/or municipality architecture shall address the entire value chain that delivers economic outcome to its primary customers—the people. It shall incorporate Federal, State, and Local Governments, public and private industries (covering facilities, hospitals, transportation, and the like) and enterprises, infrastructure ecosystem (Landscape, Transportation, Utilities, and so on), and key value stream ecosystem partners. It shall also be extensible to cover various aspects of ports of entry (roadways, airports, sea ports, and so on). Many of these are added as enterprise application stack or interface. Major considerations for smart city initiatives are energy management, green/carbon foot print, and improving the well-being of its communities.

1.2 Enterprise IOT Architecture

An enterprise IoT solution architecture spans across multiple layers. Figure 2 shows a reference enterprise IoT architecture that encompasses several layers that are device, connectivity, Aggregation (local in the form of edge gateways or edge servers), Platform (also known as IoT front end that covers event processing, storage, stream, and batch analytics), Application (purpose built), Enterprise Integration (of various system and subsystem applications, enterprise solutions), and Engagement layers (visualization of data collected in multiple forms).

Evolving cloud technologies provides us a means to design IoT applications through cloud native, microservice, and serverless based design patterns. We suggest to build IoT application as a set of discrete services that can be independently

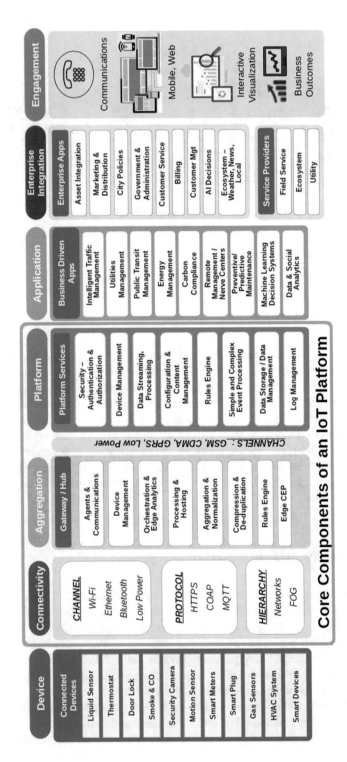

Fig. 2 IOT functional layers

deployed. We suggest to maximize the use of existing Platform as a Service (PaaS) components so that these services can scale independently. A strong architecture should provide greater scale, more flexibility in updating individual subsystems, and provide the flexibility to choose appropriate technology on a per subsystem basis. While these provide feature scale, an important factor is to have the ability to monitor individual subsystems as well as the IoT application in entirety. We recommend subsystems communicate over simple interface structure. One such interface is a human readable interface in the form of REST/HTTPS using JSON. Binary protocols should be used for high performance, secure communication requirements. The architecture should provide capabilities to cohabit hybrid cloud and distributed edge (hub, edge, fog, server) compute strategy. We see a large number of use cases where some set of data processing is happening on premise due to high data velocity, data sovereignty, regulation and compliance, and dynamic decision-making requirements. An orchestrator layer (e.g., Docker/container [2], Kubernetes [3], API Management and orchestration layers) helps to scale individual subsystems. As shown in Fig. 1, the main pillars that support the foundational layer are security, infra/service logging, and compliance.

1.3 Operating Architecture

An operating architecture covers various interconnecting functions, networks, and provides an end state view for all the agencies (enterprise, government, ecosystem players) to interface with each other on the data. It also covers the delivery, engagement models for the various smart city players. It shall provide various levels of governance, rules of engagement, and agency level data exchange needs. There are going to be multiple layers of data and multiple persona requesting this data to perform their own analysis. The operating architecture shall provide them with creating/viewing a simple or complex report or creating analytics, models (AI at City, agency, or subsystem architecture levels) for future actions. A centralized support center that can visualize all these may also be required to ensure real time tracking of the data collected (be it from sensors or from enterprise systems).

2 Typical Considerations for Each of the Layers

The IoT functional layers have differing architectural considerations, and are represented in the market by various implementations and standards, which are briefly described in the following subsections. Across each of these layers, the Smart City initiatives should also overcome several technology complexity vectors and critical components which are listed in Table 1.

Table 1 Smart City technology complexity vectors and critical components

Complexity vector	Critical components and considerations	
Interoperability and standards	Disparate closed-loop systems Limited interoperability Standards compliance and regulatory issues	Orchestration of operations Requirements (tenants, insurers, utilities, billing)
Costs and expenditures	Components (sensors, aggregation) Common IoT platform (interoperability) Connectivity (networks, interfaces)	Security (compliance, validation) Operations (IT ecosystem)
Adoption and evolution	Prioritizing use-cases Monetization of services Platforms (cloud, IoT, data, PaaS)	Legacy system integration Multi-stakeholder ecosystem
Scalability and data	System resilience and operation Volume, variety, velocity of data	Multiple sources (sensors, enterprises, integration)
Knowledge and labor force	Domain expertise End-to-end knowledge/services	Complex installation Training and re-skilling

2.1 Sensing Layer

Functional considerations involved in the Sensing Layer include implementations with direct connection to the Cloud or with interconnection via a sequence of network elements (e.g., edge, hub/router, server). Common device implementations range from single purpose sensors (e.g., temperature, humidity, Carbon monoxide, leak detection) to multi-purpose sensors that are capable of communication via local wireless links (e.g., Wi-Fi [4], Bluetooth [5], Zigbee [6], low power technologies). The market is crowded with vendors offering various implementations of such sensors. Vendors of Sensing Layer technologies include Honeywell, D-Link, Fibaro, Insteon, LeakSmart, Roost, SmartThings, Utilitech, and Texas Instruments.

2.2 Connectivity Layer

The Connectivity Layer includes short-distance as well as long-distance wireless technologies. The most common short-distance or local-area wireless solutions include such well-known standards as Wi-Fi and Bluetooth. The most common long-distance or wide-area solutions include well-known licensed technologies such as GPRS and GSM. Other evolving long-distance wireless technologies include narrowband IoT, SigFox [7], LoRA [8], and 5G Low Power Wide Area (LPWAN) networks.

2.3 Aggregation Layer

The Aggregation Layer (or Edge Layer) provides a means to accumulate, store, and/ or transport the sensor data. The implementation of this layer may include integrated sensors/hubs and gateways as well as multi-protocol multi-devices known as "integrated edge devices." Computing needs, processing needs, and non-functional capabilities like battery power and edge analytics often drive the selection of these devices.

High Performance systems may be required for "Fog layers" which converge multiple edge aggregation layers. Vendors with solutions in this space include HP, Dell, and Cisco. Devices with lower power requirements and/or devices which are purpose-built may also reside in this space. Systems including Intel Atom-based edge gateways, NXP/Qualcomm edge gateways, and mobile hubs are common design patterns. Other low power devices may be based off of increasingly popular hobbyist platforms such as Raspberry Pi [9] or Arduino [10].

2.4 Platform Layer

The Platform Layer is divided into two functional and complementary sub-layers: The Cloud Aggregation Sub-layer, and the Event and Data Management Sub-layer.

The Cloud Aggregation Sub-layer integrates events from multiple devices or sensors. Key considerations in Cloud Aggregation are scalability, security (levels of), zero/no touch orchestration, routing, processing, and quantity of data to be transferred.

The Event and Data Management Sub-layer performs minimal processing of data which is generated by devices, and stores these outputs for consumption of application/enterprise layers. Often, processing is focused on simple rules, but may include complex event processing rulesets. Key considerations for these devices are computing and scalability needs, which are driven by multiple factors including event segregation, routing, processing, and data combining. Data Management is based on short term needs for hot and warm path analytics as well as long term needs for cold path analytics, file storage, and data storage/retrieval.

Commercial/general purpose Platform Layer systems include AWS IoT [11], Azure IoT [12], Google Cloud IoT [13], ThingWorx [14], and Predix [15]. Commercial purpose-built Platform Layer systems include MachineQ [16], Sales Force IoT [17], C3 IoT [18], and SAP Leonardo [19] as well as other similar offerings. Open source Platform Layer systems include Kaa [20], ThingsBoard [21], DeviceHive [22], machinna.io [23], openHAB [24], and many others.

Typical commercial Cloud platforms such as Amazon AWS, Microsoft Azure, and Google Cloud Platform provide extensible capabilities that will drive various implementation requirements. These platforms provide large scale, multi-tenant, global deployment options, including auto scaling, full backup/restore, high availability, redundancy, and resilience capabilities. While the other platforms may provide some of these capabilities, they may not address every geography or multiple geography presence. AWS, Azure, and GCP all provide equally capable provisions in the platform and leapfrog each other in capabilities. System Integrators (for example, Cognizant) are evaluating these technologies and have put together a reference architecture framework and tool kits [25, 26].

2.5 Application Layer

While the Platform layer provides foundational element for integrating events and data management, specific work flow and end user needs are addressed via the Application Layer. Key considerations in the Application Layer include API capabilities, user interface, interfacing between web and mobile device capabilities, and other market requirements. The Application Layer provides access to the global information transport, and connects various services available in the Platform Layer. Additional important factors in the Application Layer include the development environment (SDKs, language, portability), microservices, and connectivity to other applications.

The Application Layer also covers advanced topics such as data analytics, and custom or commercial Machine Learning and AI toolkits, including Big Data tools. Commercially available tools include HDInsight [27], Caffe [28], HPE Haven [29], and Google AI [30]. In addition to several open source tools, purpose-built frameworks like TensorFlow [31], Torch [32], Spark [33], Theano [34], and NEON [35] are available to enhance and speed up application development in specific areas.

For example, Cognizant, a technology service provider to Fortune 100 companies offers a Solution platform called BigDecisions [36], which covers a wide range of ML/AI needs and is configurable for various workloads. BigDecisions platform provides a suite of BizApps for specific business processes/functions. The Data Discovery and BI Workbench solutions provide a means to drive various analytics through Selfserve Analytics & BI. The platform is built with an Infrastructure agnostic approach. It can be built with various options—Cloud, Hybrid, Multi-Tenant Hosted, and On-Premise. A rich set of SmartConnectors for faster onboarding of new information sources is already available along with configurable data Integration Workbench for industrializing data integration with automation. Comprehensive best-of-breed product partner ecosystem, Interactive Analytics Workbench, provides a means to do analytics driven Test-&-Learn sandbox for data scientists with algorithms marketplace.

2.6 Enterprise Integration Layer

In a typical enterprise, the Application Layer will need to interface with other enterprise systems to exchange information. As a result, key considerations in the Enterprise Integration Layer include technologies such as open API standards, message handling (example notifications, alerts), publish/subscribe capabilities, message bridging, message bus architecture, system management, API Management, and service identification and discovery.

Several connectors are available from commercial platforms such as Amazon AWS, Microsoft Azure, and Google Cloud Platform. Examples of notification services include Twilio, Onvoy, Spring Edge, and Nexmo. Effective enterprise systems include implementations of Customer Relationship Management (CRM), Enterprise Resource Planning (ERP), and Enterprise Asset Management (EAM). Several enterprise tools which are common in the Enterprise Integration Layer include SAP, Sales Force, Oracle, JDA, JDE, Maximo, Infor, and similar offerings.

2.7 Visualization Layer

The Visualization Layer is the engagement layer that provides user/persona Interaction with the platforms and applications. Key considerations in the Visualization Layer include customer needs and interactions. Several well-known tool sets or "UX design frameworks" are available to create the Visualization Layer, including Adobe XD, Axure RP, Balsamiq, Webflow, and Figma.

3 Security Considerations

As we scale the enterprise with sensors, interconnect, and cloud technologies, security considerations are of paramount importance at every layer and in each of the subsystems. Traditional IT/Web and/or mobile systems mostly deal with connecting people (users) to the data. In IoT systems, two new vectors come into play—sensors, distributed network of integration/aggregation points (a.k.a. gateways, Fog layers). Smart city IoT solution should address a set of necessary security principles: (a) secure provisioning of devices, (b) secure connectivity between edge devices, hubs, and fog layers, and (c) secure transmission of data between these data points and the cloud. Once the data is sent to the cloud securely, data validation, source validation (of the transmitted data), and device identity verifications along with traditional user/access mechanisms are critical to ensure secure access to the backend solutions. At every step, secure data protection via encryption during processing and storage is critical to be addressed. Several security and encryption technologies and best practices are available to tackle security management. In the USA, many of these approaches are referenced and maintained by the National Institute of Standards and Technology (NIST) [37].

Commercial cloud players [11–15] provide cloud native, PaaS based services that offer greater flexibility through a fully managed service catalog that also addresses reliable, secure bi-directional communication between IoT devices and platform services. The reference architecture should provide foundational means of secure interfacing between users, devices, and services by using per-device security credentials and access control mechanisms.

An important part of data security is data at rest. Several storage technologies are available from the cloud players that take into this critical aspect. We recommend using storage, which support encryption at rest and in motion. For user management, such as authenticating user credentials, authorization of user UI capabilities, reporting and management tools users have access to, and auditing application activities we recommend services that seamlessly provide integration between enterprise users (B2E), business users (B2B), and consumers (B2C).

Typical authentication methods make use of the "OAuth" authorization protocol [38], OpenID Connect authentication, and provide multiple levels of governance reports in the form of audit log records of system activities. Logging and monitoring for IoT application is critical in determining system uptime and troubleshooting failures. We recommend using Azure OMS and App Insights for operations monitoring, logging, and troubleshooting.

Smart city enterprises must consider failure modes and mitigation plans while implementing such large scale initiatives. High availability (HA), disaster recovery (DR) ensures that a smart city enterprise IoT system is always available, including from failures resulting from disasters to its communities. The technology used in IoT subsystems has different failover and cross-region support characteristics be it an IoT device failure, network failure, and/or security breach/concern. Smart city enterprise IoT applications must consider hosting of duplicate services and where appropriate duplicating application data across multiple regions. Smart city applications need to address certain service levels such as failover downtime and data loss characteristics. Using these requirements, a resilient system needs to be designed.

4 Interoperability

A major aspect for smart city enterprise IoT systems is tackling interoperability among all the constituent systems. As shown in Fig. 2, enterprise systems and integration layers provide several data exchange mechanisms. Several programs are underway and organizations are in the process of creating and proposing architectural design principles, taxonomies, and standards. The process of standards convergence is critical to enable IoT's full potential to be realized in the smart city market. While this is a nascent subject, the U.S. Department of Commerce's National Institute of Standards and Technology (NIST) is bringing together several domestic and international partners. It has launched the "International Technical Working Group on IoT-Enabled Smart City Framework." This is now shortened to "IES-City Framework" which is pronounced "yes city" framework [39, 40].

5 Conclusions and Final Thoughts

This chapter has presented and discussed several aspects of IOT architecture which are important for Smart City implementations. This technology area is evolving and expanding at a very quick pace. As a result, the overview coverage here is necessarily brief and does not encompass all aspects. However, the basic concepts and conceptual architecture framework presented are robust, and should provide the reader with a useful starting point for exploring the constituent technologies, standards, and implementation issues associated with IOT architecture in a Smart City environment.

References

1. IDC's worldwide digital transformation use case taxonomy, 2018: smart cities and communities (#US44261515, September 2018)
2. Docker. https://www.docker.com/
3. Kubernetes. https://kubernetes.io/
4. Wi-Fi Alliance. https://www.wi-fi.org/
5. Bluetooth Special Interest Group. https://www.bluetooth.com/
6. Zigbee Alliance. https://www.zigbee.org/
7. Sigfox. https://www.sigfox.com/
8. LoRa Alliance. https://lora-alliance.org/
9. Raspberry Pi Foundation. https://www.raspberrypi.org/
10. Arduino AG. https://www.arduino.cc/
11. Amazon Web Services – IOT. https://aws.amazon.com/iot/
12. Microsoft Azure IOT. https://azure.microsoft.com/en-us/overview/iot/
13. Google Cloud IOT. https://cloud.google.com/iot-core/
14. ThingWorx Industrial IOT. https://www.ptc.com/en/products/iot/thingworx-platform
15. GE Predix Platform. https://www.ge.com/digital/iiot-platform
16. Comcast MachineQ. https://machineq.com
17. Salesforce IOT. https://www.salesforce.com/products/salesforce-iot/overview/
18. C3 Enterprise AI. https://c3.ai/
19. SAP Leonardo. https://www.sap.com/products/leonardo.html
20. Kaa Enterprise IOT Platform. https://www.kaaproject.org/
21. ThingsBoard Open Source IOT Platform. https://thingsboard.io/
22. DeviceHive Open Source IOT Data Platform. https://devicehive.com/
23. Applied Informatics. macchina.io. https://macchina.io/
24. Open Home Automation Bus. https://www.openhab.org/
25. Cognizant (2019) Solution overview: accelerating IoT programs at scale: connected reference architecture & toolkit. https://www.cognizant.com/Resources/connected-reference-architecture-and-toolkit-overview-hires.pdf
26. Cognizant (2019) The five essential IoT requirements and how to achieve them. https://www.cognizant.com/whitepapers/the-five-essential-iot-requirements-and-how-to-achieve-them-codex4241.pdf
27. Microsoft HDInsight. https://azure.microsoft.com/en-us/services/hdinsight/
28. Caffe. Berkeley AI research. http://caffe.berkeleyvision.org/
29. Hewlett Packard Enterprise. Haven on demand. https://www.havenondemand.com/
30. Google AI. https://ai.google/

31. Google TensorFlow. https://www.tensorflow.org/
32. Torch: a scientific computing framework for LUAJIT. http://torch.ch/
33. Apache Spark: unified analytics engine. https://spark.apache.org/
34. Univ. Montreal MILA Lab: Theano. http://deeplearning.net/software/theano/
35. The National Ecological Observatory Network. https://www.neonscience.org/
36. Cognizant BigDecisions Platform. https://www.cognizant.com/bigdecisions
37. National Institute of Standards and Technology (NIST). https://www.nist.gov/
38. Hardt D (ed) (2017 Oct) RFC 8252, "OAuth 2.0 for Native Apps." https://tools.ietf.org/html/rfc8252
39. National Institute of Standards and Technology (NIST), Engineering Laboratory. Cyber-physical systems, IES cities architecture. https://www.nist.gov/el/cyber-physical-systems/ies-cities-architecture
40. National Institute of Standards and Technology (NIST). International technical working group on IoT-enabled smart city framework. https://pages.nist.gov/smartcitiesarchitecture/

Measuring Innovation: Tracking the Growth of Smart City Ideas

Steve Pearson

Contents

1 Introduction

Where is the Smart City headed? And how quickly will we get there?

To help in understanding the trajectory of innovation in "Smart Cities," some formalized analysis is necessary and useful, particularly for inventors, strategists, and business practitioners.

Spoiler alert: Innovation continues a strong upward swing and my investigation points out leaders behind that uptick. Skip to Data/Results if you must, but stick with me for methods and explanation.

Let's step through the process together.

Many people get excited about innovations (me too), but seldom is a new idea the next "new" thing reaching the marketplace. That's because what gets publicized as significant innovation is often based only on opinion (albeit some are from "experts") and what makes juicy click bait. Worse yet, these predictions are usually not scientific as they're not measurable, reproducible, or unbiased.

What are some possible ways to analyze individual innovations occurring in the Smart City space within scientific parameters?

S. Pearson (✉)
The Pearson Strategy Group, LLC, Austin, TX, USA
e-mail: steve@pearsonstrategy.com

© Springer Nature Switzerland AG 2020
S. McClellan (ed.), *Smart Cities in Application*,
https://doi.org/10.1007/978-3-030-19396-6_9

Perhaps the best way is to gather a broad swath of innovation-related data.

One option is to do an analysis of the marketplace, but the results would be soon outdated as today's products have gone through years of research and development and will soon be innovation has-beens.

Another option is to analyze the various technologies that are being researched, but this method would suffer from not being analytical. It would also likely be skewed by the reviewer making choices on what to include and exclude, and it would likely be skewed toward the academic perspective, not the future implementation perspective.

These two methods are common, but we don't include them here as they are not measurable, reproducible, or unbiased.

My choice for this analysis is worldwide patenting activity.[1]

2   Patents: An Information Treasure Trove

Patent applications and issued patents ("granted patents") are legal documents and forms of Intellectual Property (IP). These documents are regulated by the laws and procedures of each country's patenting authority, although some countries will band together to form wider ranging patenting authorities. The European Patent Office [1] is one of the better-known examples.

Many people think that their market/product awareness reflects the contents of IP filings. In reality, patents include a much larger set of data that only mildly reflects the current marketplace. This difference is significant since "80% of the information found in patents is not found elsewhere" [2].

Similarly, engineers and technical experts often think their individual knowledge, combined with the collective knowledge of others in published research papers, makes them a great measure of innovation. In fact, "patent databases are more robust and provide a more accurate picture of the state of the art in the field of interest," because "the researcher decides which references to cite in a paper." This is in sharp contrast to patents where the inventor is required "to cite all patents used to develop an innovation even if they belong to a competitor" [3].

Further, the relevant citations must include the results identified by the patent searcher and those of the patent examiner, who works at the patenting authority and accesses the novelty of the submitted patent application. The intended result of this process is that only those ideas that are unique become patents. Note that the standard for what becomes a patent is beyond the scope of this chapter since this is a complex subject dependent upon the many rules in place at each patenting authority.

The result is that the owner of the idea will be provided a competitive advantage if his patent is issued in exchange for disclosing his idea. It is this hope that drives many people and companies to submit their ideas in patent applications as early as

---

[1] Please note that the information in this chapter does not provide investment or legal advice.

possible to prevent someone else from beating them on a patent on the same idea. This timeliness helps to explain why many countries formed patent systems. They spur their economies by growing innovation by requiring that innovators disclose their inventions in detail so that others may learn from them and continue making improvements.

Many consider patent applications and granted patents as strictly legal documents. However, they also provide a bounty of non-legal information known as "Competitive Intelligence."

Some of the more obvious insights are learning about the inventor(s), owner(s), and technologies behind each innovative idea. But patents also afford us a look into less obvious relationships between the legal, marketplace, and technology worlds we often keep disconnected. With this knowledge we're able to derive several benefits, including:

- Identifying differing technologies that are resolving the same problem.
- Discovering which companies and inventors are behind innovation clusters.
- Determining whether a technology is in its infancy or late in life.
- Recognizing which technologies are seeing significant investments.

It's important to note that many patent applications remain unpublished for 18 months after filing by many of the world's patenting authorities, including the USA. This "lag time" between filing and publication will become important later as we develop a trend line of innovation activity.

The legal protection offered by patents can be quite compelling and, accordingly, about 4.5 million patent-related publications are published annually worldwide. This number is growing by roughly 10% per year [4].

Investing in the opportunity to get a utility patent, what many people consider to be the "typical" type of patent, is not for the faint of heart as costs often start at $10,000 in the USA but are often much higher in other locations. The word "opportunity" is used deliberately as roughly 50% of patent applications received at the USPTO turn into issued patents. Plus, most inventors don't have a clue about whether their idea will grant since they did no research to determine how innovative their idea was, before it was filed. The very high-level flowchart shown in Fig. 1 provides a more reasoned approach to investing and preparing for making money than jumping in too quickly and paying for a worthless patent.

These significant costs form a "barrier to entry" for some inventors and companies but, for our purposes, they help to ensure that the inventor/assignee thinks that their idea is potentially novel and has marketable value. There is a lot of novelty that is unique but has little commercial value and will not need patent protection, but many applicants don't know this at the time of filing. If you think that the process described in Fig. 1 is not important, consider the commonly cited statistic that 95% of granted patents are never monetized!

In fact, investments made in research and development and patents can project trends as companies are investing in developing novel technologies and protecting them in a future marketplace.

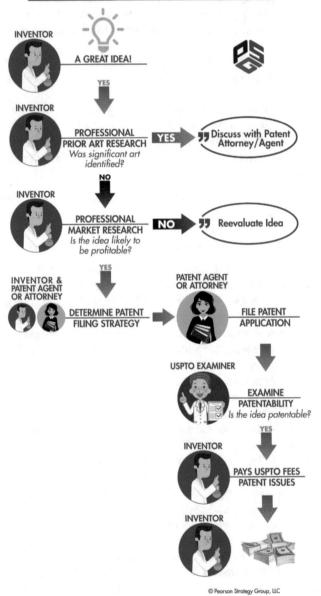

Fig. 1 High-level flowchart of the patent process

3 Patent Analytics: Data Mining Patent Publications

With the 4.5 million annual patent publications published annually worldwide, the set of historical applications and issued (or "granted") patents create a rich and statistically reliable source for this research.

This large number, combined with a great amount of detail in each, allows for the statistical analysis of even highly specialized areas of technology using a branch of research termed "Patent Analytics" (or "Patent Landscapes"). It's also worth mentioning that it's important to use the worldwide resources since only 19% of the patent applications are being filed in the USA, see Fig. 2 [4].

Patent Analytics is perhaps the fastest and most cost-effective, data driven predictor of future marketplaces, at least those in technology spaces. Benefits include:

- Affords early risk and opportunity identification giving companies more time to react to future marketplace changes.
- Allows companies to refine their IP and company strategies for maximum advantage.
- Accelerates the R&D cycle and time to market.
- Minimizes costs by avoiding pointless R&D and patents.

In short, we're going to leverage the benefits of Patent Analytics to generate a numerical assessment on what people and companies are filing as patent applications thinking that their idea is innovative.

Note that there are many strategies for how and when to file patent applications and that these discussions are beyond the scope of this chapter; not all patent filings are made solely to gain a valuable patent. While some of the companies filing patent

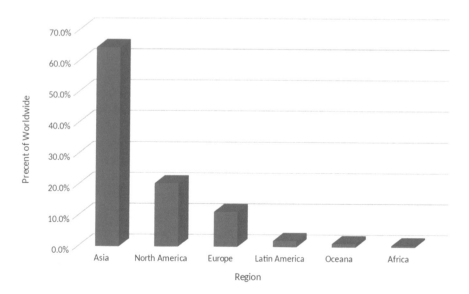

Fig. 2 Patent applications filed by world region

applications are very knowledgeable in their area and have a good grasp of what is innovative, many other companies and individual inventors are not. This will not be a problem for this analysis as the statistical methods used will reduce less innovative ideas into the noise of the data with minimal impact on the more statistically relevant data we're seeking.

There are varying procedures and laws in place throughout the many patenting jurisdictions around the world, currently around 190 countries [5, 6]. These variations include whether patent applications will be published or not, when the issued patent will be published and so on. This chapter will not account for these variations as the volume of data will often make these differences negligible.

Patent-related data can reveal new or hidden (from the marketplace) entrants into a space that might otherwise be missed. Examples include research organizations, Non-Practicing Entities (NPEs), and companies such as ARM Limited [7] which develop new technologies and IP with a focus on licensing them to other companies.

To help focus on measuring innovation: most of the data presented below will be shown as "patent families." Per the European Patent Office (EPO), a simple patent family "is a collection of patent documents that are considered to cover a single invention. The technical content covered by the applications is considered to be identical" [8]. In other words, all the issued patents and their associated patent applications and resulting issued patents related by technology, date, and inventor can be lumped into a single family. Some families are very simple and contain a single patent application, while others can contain hundreds of very similar patent applications and issued patents from various patenting authorities around the world.

As an example of one of the many patent families identified in this research, Table 1 is a single patent family containing seven publications belonging to Samsung Electronics for "Electronic Device and Control Method Thereof."

By using patent families, we minimize the repetitive information contained in multiple applications and patents and shift our focus from the number of publications to the number of innovations.

One of the statistical results that will be created is a trend line showing how patent families are progressing over time. This will be useful as we'll be able to ascertain where Smart City innovations are in the technology adoption lifecycle, as shown in Fig. 3.

Table 1 Patent family example (Samsung Electronics)

Authority	Document type	Publication number	Publication date	Application date
China	Patent Application	CN108353361 A	20180731	20161027
Europe	Patent Application	EP3335479 A1	20180620	20161027
Europe	Patent Application	EP3335479 A4	20181107	20161027
Korea	Patent Application	KR20170048839 A	20170510	20151027
U.S.	Patent Application	US2017118323 AA	20170427	20161027
U.S.	Issued Patent	US10009452 BB	20180626	20161027
WIPO	Patent Application	WO17074045 A1	20170504	20161027

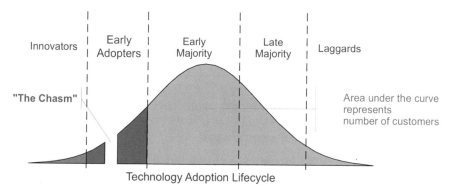

Fig. 3 Technology adoption lifecycle [9]

Companies and inventors generally know that they will usually be wasting their time and money to continue investing in R&D for well-adopted technologies in saturated marketplaces which are being represented on the far-right side of the curve in Fig. 3.

The patenting associated with the technology adoption lifecycle curve also follows a normal distribution/bell curve similar to Fig. 3. This curve's timeliness is generally ahead of that of the technology adoption lifecycle since investment and research will often decline as market adoption increases (assuming no other significant changes are occurring).

Finally, it is important that the numbers revealed by patent analytics not be used for numerical comparisons as that would be beyond the scope of what these numbers are showing. Use only qualitative comparisons.

Patent Analytics reveals a lot of helpful data but is not comprehensive. This data should not be used for legal and financial decisions in the absence of other market, technical, legal, or other sources. Patent Analytics answers many questions, then empowers the reader to ask more sophisticated questions.

4 Methodologies: Zeroing in on Smart City Terms

One aspect of using Patent Analytics to do research is that we should identify the data with a method that's as impartial and comprehensive as possible. This step will help bring out data we may not have anticipated.

The very definition of "Smart City" pushes the realm of this research in the direction of high-level Information Technologies (IT) and communications capabilities. "Smart Cities" function or add cross-functional value to the many siloed technologies which could, generally, function on their own and independent of the many other technologies. A "Smart City" could be described as System of Systems, while the technologies that underlie making a city "smart" include everyday items such as street lighting, but also includes more recent innovations such as the Internet of

Things (IOT), fifth generation of cellular mobile communications (5G), intelligent energy delivery (smart grid), and so on.

Three overarching methods for finding relevant Smart City patent families are:

1. Search for keywords (English and non-English are generally preferable for international research like this).
2. Search using one or more of the various patent classification codes in the various classification systems.
3. Some combination of the first two methods.

Unfortunately, Smart Cities encompass a wide breadth of technologies and applications. This is similar to the Internet of Things (IOT) in that the fundamental technologies have generally been around awhile and encompass many applications but have been collected under a common heading in recent years. "Smart City" can be included unequally since it's an ambiguous phrase that can be applied, or not, according to the author's preferences and language. The goal will be to assemble one or more strategies that collect most or all of the Smart City patent families of interest while controlling the amount of irrelevant IP which could distort the data analysis.

Also, we should consider whether we're trying to include or exclude Smart City functionality, planning/design, or operational technologies. We will not focus on bringing out any of these segments to the exclusion of the others.

A review of patent classification codes found that neither the International Patent Classification (IPC) [10] nor the Cooperative Patent Classification (CPC) [11] includes any Smart City specific codes. This isn't unusual for newer technologies as codes will be described and implemented only once a reasonable number of patent and applications fit into a needed category.

Since the patent classification systems don't contain any relevant codes, we'll need to focus on text searches. We could simply search for every occurrence of "Smart City" but this would be so restrictive that we would miss out on what could be large datasets. We should, therefore, consider all known synonyms as well as considering:

1. Companies known to be operating, researching, investing in the market space.
2. Mention made to Smart City standards (e.g., International Organization for Standardization (ISO) [12], Institute of Electrical and Electronics Engineers (IEEE) [13], etc.).
3. Review technical publications to find additional terms.
4. Review non-technical publications to find additional terms.
5. Using Patent Analytics to help detect significant information which can be iteratively used to improve and refine our search strategies.

Standards can be helpful for this type of research since they cover established technologies that may be described differently from person to person. Some standards are very narrow such as the "thousands of electrical standards exist today" [14].

Table 2 Smart City dataset

Patent families	7456
Patenting jurisdictions	72
Patent applications	20,404
Issued patents	6587

5 Data/Results

The use of the above methodologies resulted in a Smart City dataset shown in Table 2.

The earliest non-translated usage of "Smart City" or "Smart Cities" in a patent-related publication is US2010228601 [15] which was published on Sept. 9, 2010, only eight years before the initial research for this article. As you'll see below, the usage of this terminology has expanded significantly since then. And by the way, the Smart City dataset described above encompassed significantly more terminology!

6 Analysis

Using patent families as the focus of innovation, we can see from the orange line in Fig. 4 that innovations in Smart City are rising dramatically with no sign of slowing down. This likely means that we're still early on the Technology Lifecycle Curve with plenty of room to grow.

As mentioned earlier, patent application publications are generally delayed by 18 months, so we deliberately did not seek out these after 2016 to prevent inducing a misleading drop-off in trend line activity. Further, this data was generated before the end of the year (2018) which will cause the number of patents and families to be slightly lower.

Like the number of families, the number of patent applications shown by the blue line is also growing strongly, Fig. 4. This indicates that a significant amount of R&D and investment in IP is being made by the filers to protect their ideas in the potentially lucrative marketplace of the future.

Note that patents are shown in Fig. 4 by the gray line but there's an even longer lag time to publication than that of applications, estimate three to five years based on what's typical in the USA; the number of patents being much lower than applications is expected.

Breaking down the assignees (the owners of the patents and applications) in Fig. 5 we can see the top three companies with the most families are well-known in the electronics and communications industries: Samsung, Cisco, and AT&T.

Samsung is the clear winner with the largest number of Smart City patent families, see Fig. 5, and the largest volume of applications. Samsung's level of innovation is higher than its next two highest competitors, Cisco and AT&T; and, more interestingly, all nine other companies combined! Don't forget that the substantial

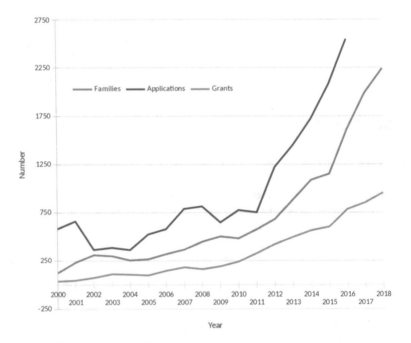

Fig. 4 Smart City families and publications by year

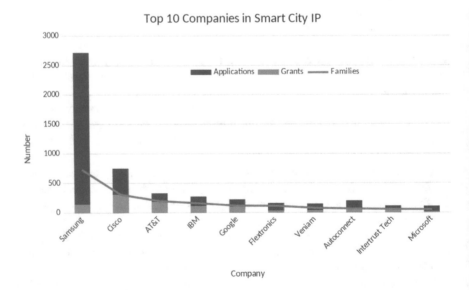

Fig. 5 Top 10 companies in Smart City IP

cost of generating, filing, and obtaining patents means Samsung is investing a significant amount of money to protect its (also costly) R&D activities.

It should be noted that Samsung's issued patents tally is lower than Cisco and AT&T, but unlikely to be important. About half of U.S. patent applications can be expected to eventually become U.S. issued patents, with some variance expected for other countries [16].

Another explanation for Samsung's lower issued patents total is that the lag time for receiving issued patents is long in comparison to the short periods of time when most of its patent applications have been filed, see Fig. 6. The graph for the five most active companies is labeled "velocity" to indicate the rates of new families and publications as well as the rates of change for those same quantities.

The family counts in Fig. 6 show that Samsung is rapidly picking up the pace in investing in Smart Cities IP and that its overall numbers will likely continue to grow in coming years. Conversely, Cisco and AT&T are not showing any appreciable changes in their new patent families and, therefore, are not likely to be nearly as innovative as Samsung in the next few years.

Taken together, Samsung's large investments into R&D and future patents indicate that it sees a large opportunity in Smart Cities. Of course, this conclusion is made without considering non-patent data.

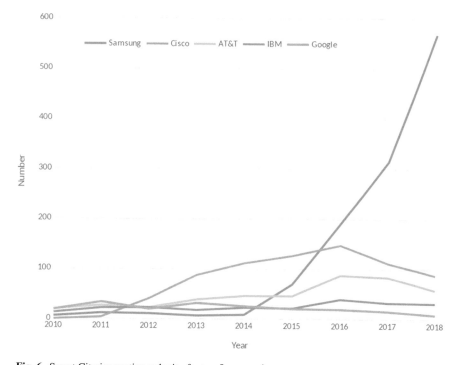

Fig. 6 Smart City innovation velocity for top 5 companies

Table 3 Description of areas examined in Fig. 7

Area	Scope
Sensors	Sensors, Internet of Things
Wireless	Wireless communications
Cameras, Video	Cameras, video, pictures
Money	Currency, electronic payments, shopping
Vehicles	Automotive vehicles, roadways, traffic control
Health	Health, medical, fitness
5G	Fifth generation of cellular mobile communications
Global warming	Greenhouse gases
Energy	Energy generation, transmission, or distribution
AR, VR	Augmented reality, virtual reality, mixed reality
System intelligence	Artificial intelligence, machine learning, predictive algorithms
Food, Restaurants	Food, food equipment, groceries, restaurants
Privacy	Privacy, surveillance, facial recognition

So what fields of science are these patents and applications identifying as future market opportunities? A macro-analysis of several areas described in Table 3 is shown in Fig. 7.

Which parts of the world are seeing the most activity? The five most active patent jurisdictions are shown in Fig. 8. Do note that the requirements for applications vary by jurisdiction and that the USA is often number one or two for many technology areas as compared to China.

7 Updates About Smart Cities Innovations Available

The Pearson Strategy Group [17] is a key provider of patent searches, patent analytics, market research, and competitive intelligence. Businesses and inventors rely on PSG's expertise in Intellectual Property, technology, and market assessment to reduce costs and minimize risks while accelerating the innovation cycle.

As part of these strategic services, the group routinely provides industry updates and finely tuned custom reports that are cost-effective, efficient, and valuable for decision-makers.

Now, to help businesses, entrepreneurs, inventors, and city officials maintain their knowledge on Smart Cities innovations, the Pearson Strategy Group provides updates on global activity.

PSG clients often use this type of research to:

- Identify emerging technologies or processes.
- Leverage disruptive technologies.
- Evaluate patentable ideas for utility and differentiation.
- Detect patent opportunities and threats in near real-time.

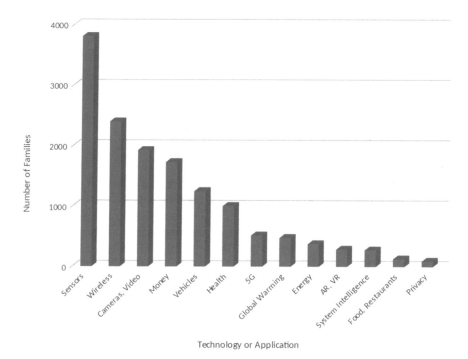

Fig. 7 Smart City technology and application analysis

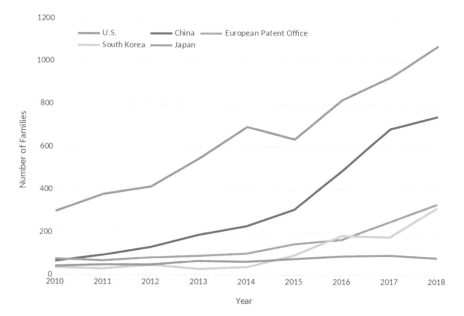

Fig. 8 Smart City innovation velocity for Top 5 jurisdictions

- Minimize risk of patent infringement.
- Maximize IP protection options.
- Empower legal, R&D, and engineering teams with facts, not hunches.
- Generate new ideas on how to improve their products.
- Stay one step ahead of their competition.

8 Conclusion

The many terms used to identify the dataset for this analysis revealed a lot of issued patents and patent applications. However, the analysis can't be considered complete. In fact, it is the filer's and their patent agent/attorney's decision on whether to use these terms. Realistically, many of these technologies have likely been around awhile and are only now being "tagged" with terms specifically associated with Smart City. This is one of many reasons why the above analysis is not intended to provide legal or financial advice.

References

1. European Patent Office - Home Page. European Patent Office [Online]. https://www.epo.org/
2. Trippe A (2016) Guidelines for preparing patent landscape reports. World Intellectual Property Organization (WIPO)
3. Simões J (2018) Patent mining indicates promising routes for research. Phys.org
4. World Intellectual Property Organization (WIPO) (2017) World intellectual property indicators 2017
5. World Intellectual Property Organization (2017) The PCT now has 152 contracting states. World Intellectual Property Organization
6. Hoffman LJ (2015) Countries in which the Patent Cooperation Treaty (PCT) does NOT apply, Hoffman Patent Firm
7. C. U. Arm Limited. ARM. Arm limited [Online]. https://www.arm.com/
8. European Patent Office. DOCDB simple patent family. European Patent Office
9. Chelius C (2009) Technology adoption lifecycle. Wikimedia [Online]. https://commons.wikimedia.org/wiki/File:Technology-Adoption-Lifecycle.png
10. World Intellectual Property Organization. International patent classification (IPC) [Online]. https://www.wipo.int/classifications/ipc/en/
11. Cooperative Patent Classification. About CPC. USPTO, EPO [Online]. http://www.cooperativepatentclassification.org/about.html
12. International Organization for Standardization. International Organization for Standardization [Online]. https://www.iso.org/home.html
13. Institute of Electrical and Electronics Engineers. IEEE. Institute of Electrical and Electronics Engineers [Online]. https://www.ieee.org
14. McClellan S et al (eds) (2018) Smart Cities: applications, technologies, standards, and driving factors. Springer, pp VII, 239. ISBN 978-3-319-59381-4. https://www.springer.com/us/book/9783319593807
15. Raj Vaswani SMF (2010) Method and system of applying environmental incentives. US Patent US20100228601A1, 9 Sept 2010
16. U.S. Patent Statistics Chart, Calendar Years 1963 - 2015. U.S. Patent and Trademark Office, 11 February 2019 [Online]. https://www.uspto.gov/web/offices/ac/ido/oeip/taf/us_stat.htm
17. Pearson Strategy Group. https://pearsonstrategy.com/smart-city-innovation

Index

© Springer Nature Switzerland AG 2020
S. McClellan (ed.), *Smart Cities in Application*,
https://doi.org/10.1007/978-3-030-19396-6

Printed in the United States
By Bookmasters